WORKING IN
JAPAN

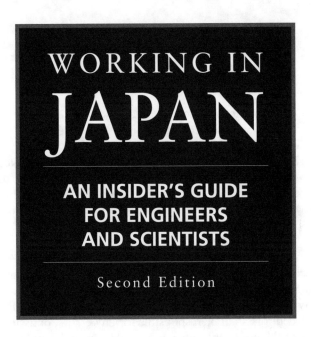

WORKING IN JAPAN

AN INSIDER'S GUIDE FOR ENGINEERS AND SCIENTISTS

Second Edition

HIROSHI HONDA

Editor/Author

ASME PRESS

New York 2000

© 2000 by The American Society of Mechanical Engineers
Three Park Avenue, New York, NY 10016

Library of Congress Cataloging-in-Publication Data

Working in Japan : an insider's guide for engineers and scientists / edited by
  Hiroshi Honda.—2nd ed.
    p.   cm.
  ISBN 0-7918-0152-7
  1. Engineering—Japan. 2. Engineers—Employment—Japan. 3. Japan—
Social life and customs. I. Honda Hiroshi
TA157.W68 2000
331.12′92′000952—dc21   00-028390

Contents

PART 5   THE BENEFITS OF PROFESSIONAL SOCIETIES   141

PART 6   EMPLOYMENT CASE STUDIES   165

Foreword

JAPAN AND THE GLOBAL SOCIETY _____

At this turn of the millennium, many things have changed in Japan since the first edition of this book was published in 1991. Foreign-born professionals are no longer strangers in Japanese society, and increasing numbers of their Japanese colleagues and hosts have now become accustomed to dealing with them. Various personal gaps between foreign-born professionals and their Japanese counterparts and hosts have been narrowed, even though there still are substantial language barriers and cultural gaps.

The most significant change in recent years in the hiring of foreign nationals in Japan is a significant increase in postdoctoral fellowship programs, half of whose 10,000 positions are now allowed to be filled with qualified foreign engineers and scientists. Universities are gradually increasing their modest number of faculty positions for foreign-born engineers and scientists, and companies have constant needs for foreign-born professionals.

Japanese society needs the foreign-born professionals to cope with progressive global and local standards and with changes in such areas as rules and customs for commercial dealings, engineering practice, and scientific work, which have come about with the globalization of society and of many professions. A significant proportion of foreign-born professionals come to love the Japanese people and their culture and society. Increasing numbers of them are staying in Japan longer than ever before, and some even come back to Japan after having to leave the country for some time. However, some foreign-born professionals never come back again to Japan. It is likely that more would, if they were given proper guidance in comfortably pursuing their professional work and enjoying everyday life in the country.

GLOBALIZATION OF THE ENGINEERING PROFESSION _____

We have been discussing the subject of globalization at meetings of the ASME Industry Advisory Board (IAB) and at other ASME meetings ever since we held the first IAB panel session on the globalization of the engineering profession on November 10 and 11, 1993, in Washington, D.C. Normally, when confronted with a challenge, engineers rely on their technical proficiency and ingenuity. They have built small but powerful computers, reusable space vehicles, and a tunnel under the English Channel—often after skeptics said it could not be done. Today, however, engineers face a new challenge with no convenient technical formula: globalization. How engineering educators respond will determine, to a great extent, how good a job future engineers do in meeting this challenge.

Few would contest that the rapid spread of free markets, combined with the "shrinking" of the world through telecommunications, is having an increasingly pronounced effect on the engineering profession. Increased global competition requires engineers to couple technical innovations with sensitivity to the world marketplace. Global telecommunications open the door for engineers to work at home or abroad. And megamergers, such as the union of Daimler-Benz and Chrysler, and alliances such as Ford and Mazda or Microsoft and Softbank (headquartered in Japan), prompt corporations to bring engineers together across national and cultural boundaries. The question for engineers is an obvious one: What is the best preparation for succeeding in a dynamic, increasingly international workplace?

The short answer is that engineers must strive to be able to work anytime and anywhere. Engineers helped bring about the era of globalization by designing and building the computers, satellites, telecommunications equipment, and Internet that are its backbone. Now they must respond to the global challenge they have enabled.

REMARKS ABOUT THIS BOOK _____

The rapid pace of change challenges all authors of books and articles to convey a message that will stay timely. This book provides a guide to future global opportunity and a sound foundation for those who include Japan as a part of their vision.

Those who work in the business and industry sectors may be interested in a sister book of this volume, entitled *Working in Japan: An Insider's Guide for Business and Industry Professionals.*

From the Authors of the foreword

I feel pleased and honored to be in a position to continue to pursue this work, and with a great satisfaction, I would like to offer this volume to the public at this turn of the millennium.

Hiroshi Honda
Editor/Author

It is my pleasure on behalf of ASME to acknowledge the contributions of Dr. Hiroshi Honda to the engineering community through this timely second edition.

Winfred M. Phillips
President, ASME, 1998–99
University of Florida

Preface

This book is intended to help the significant number of foreign-born professionals arriving in Japan to cope with living and working in their new cultural environment. Since many of these professionals initially have a limited knowledge of Japanese culture and language, they often encounter difficulties with Japanese mannerisms, customs, and patterns in human relations.

Foreign-born professionals are, of course, expected to bring with them some of the uniqueness and beauty of their own culture, which Japanese people do appreciate. However, there are unwritten rules, which the Japanese people have developed over a long period of history for the smooth and efficient functioning of their society. If people from other backgrounds are aware of these unwritten rules and are able to adapt to them, they can enjoy a pleasant and fruitful life in Japan, both professionally and privately.

Fortunately, among the members of ASME (the American Society of Mechanical Engineers) International and their friends, we find people of many nationalities and cultural backgrounds who have had experiences in Japanese industry or at Japanese national institutes or Japanese universities and colleges. Their cross-cultural experiences and their insights into Japanese society are collected and set forth in this book, which should serve as a convenient and useful guide for the newcomer to Japanese society.

The book addresses the concerns of foreign-born engineering professionals and scientists in a logical sequence. Part 1, the Introduction, surveys the current employment of foreign professionals, including engineers and scientists in Japan, as well as future trends. Part 2, Applying for a Job in Japan, introduces effective methods of applying for a position in industry or at a national institute or a university or college, and describes typical salary and benefit structures in Japan. Part 3, Facets of Japanese Society, traces the history and describes the features of Japanese society, comparing it with societies in western and other Asian cultures. Part 4, Cultural Gaps and the Language Barrier, summarizes some of the problems of adjustment experienced by western professionals. Part 5, The Benefits of Professional Societies, describes the opportunities afforded by involvement in the activities of international societies such as ASME International, the International Congress of Mechanical Engineering Societies (ICOMES), European societies, Asian societies, and Japanese professional societies and organizations. Part 5 also covers the trend toward mutual recognition of engineering and technology education and qualification among the United States, Japan, and other nations, and the role and efforts of the Accreditation Board for Engineering and Technology (ABET), U.S.A., and the Japan Accreditation Board for Engineering Education (JABEE). Part 6, Employment Case Studies, recounts the experiences of professionals of different nationalities who have worked in Japanese industry, at a national institute, and at Japanese universities, and of some who have established a firm in Japan. Part 7, Conclusions, first describes both the opportunities and the issues facing foreign-born professionals in advancing their careers after returning from or recontinuing employment in Japan. Practical information on Japanese hiring practices is then offered, based on replies given by a number of Japanese companies to an editor's survey conducted at this turn of the millennium.

I hope this book will be a valuable reference for foreign-born engineers and scientists who are interested in working in Japan or are already planning to come to Japan to work.

Hiroshi Honda

Acknowledgments

I would like to thank Mary Grace Stefanchik, Acquisitions Editor, Philip Di Vietro, Director, Technical Publishing, Tara Smith, Production Coordinator, Jeff Howitt, Director of Marketing and their staff at ASME Press for their support, without which this second edition would not have been completed. I would also like to thank Dr. David Belden, Executive Director, Mr. David Soukup, Managing Director, and Mr. Arnold Rothstein, Vice President for Engineering and Technology Management Group of the American Society of Mechanical Engineers International for their backup support in many areas.

I am honored and privileged that Dr. Winfred Phillips has accepted my invitation to contribute to this book. In addition, I would like to thank Dr. Peter E. D. Morgan and other authors for updating their original contributions to the first edition for inclusion in this new edition and for recruiting new contributors to this volume.

Dr. Robert Latorre, Dr. George Peterson, Ms. Kathryn Aberle, Dr. Joyce Yamamoto, Dr. Giuseppe Pezzotti, and Dr. Junjiro Iwamoto are new contributors to this edition; their contributions have enhanced the value of this volume. I have to also note that Dr. Ichiro Watanabe, Dr. Kazuo Takaiwa, and Mrs. Deborah Coleman Hann, contributors to the first edition, have passed away and would like to share my deepest sympathies on their passing.

Mr. Raymond C. Vonderau, Dr. Kazuo Takaiwa, Mr. Daniel K. Day and Dr. Shuichi Fukuda served as contributing editors for the first edition of *Working in Japan: An Insider's Guide for Engineers*, without which this second edition would not have been brought about.

Last, I fondly remember the support that Dr. Arthur E. Bergles, President of ASME (1990–91), and Dr. Wataru Nakayama, Chairman of ASME JAPAN (1988–92), extended to us toward publication of the first edition of *Working in Japan: An Insider's Guide for Engineers.*

I gratefully acknowledge all this help and the contributions of all the authors, without which this book would not exist.

Hiroshi Honda

PART

1

Introduction

1

Employment of Foreign-Born Professionals in Japan: Background of the Current Trend

Hiroshi Honda

INTRODUCTION

Japanese companies and various other organizations in Japan came to hire significant numbers of foreign-born professionals[1] in the 1990s, after the Japanese economy grew at a high rate and many Japanese organizations became internationalized during the period from 1985 to 1991. As Figure 1.1 shows, the number of incoming foreign-born professionals had increased from 43,994 in 1985 to a peak of 113,559 in 1991, when the economic bubble in Japan collapsed. The number decreased from 1991 to 1993, with a jump in 1994, and again decreased to 81,508 in 1995. On the other hand, the number of the foreign-born professionals *staying* in Japan for work steadily increased from 20,478 in 1984 to 105,616 in 1994, and then decreased in 1995, as shown in Figure 1.2. This number has steadily increased again from 87,996 in 1995 to 118,996 in 1998, even though Japan's real GDP growth rates were 0.8 percent, 3.5 percent, 1.5 percent, and –2.0 percent from 1995 to 1998. The rate of Japanese jobless workers,

---

[1] The phrase *foreign-born professionals* refers to foreigners who are legally allowed to work in Japan because they are qualified in any of the following 14 occupations: professor, artist, religionist, journalist, investor/business manager, legal/accounting professional, medical service professional, researcher, (language) instructor, engineer, specialist in humanities/international services, intracompany transferee, entertainer, and skilled labor. (Ministry of Justice, Japan, Immigration Control and Refugee Recognition Act.)

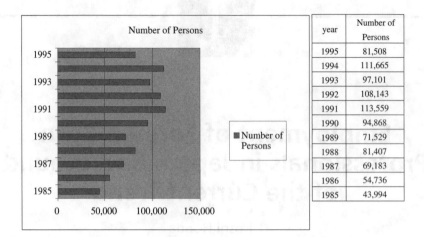

Source: Statistics of Administration of Arriving and Departing Foreign Nationals, Ministry of Justice, Japan

Figure 1.1   Number of incoming foreign professionals arriving for work in Japan, 1985 through 1995. (From Ministry of Justice, Japan, *Statistics of Administration of Arriving and Departing Foreign Nationals*.)

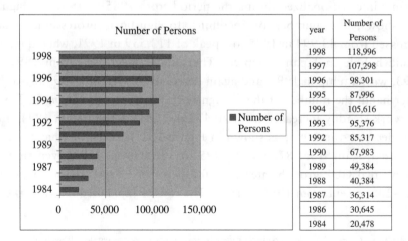

Source: Statistics of Foreign Nationals' Residence, Ministry of Justice, Japan

Figure 1.2   Number of foreign professionals staying for work in Japan, 1985 through 1998. (From Ministry of Justice, Japan, *Statistics of Foreign Nationals's Residence*.)

a majority of them ranging from middle age to over 50 years old, reached a peak of 4.8 percent as of March 1999, resulting from drastic restructuring of industries.

It was only in the first quarter of 1999, that is, January to March 1999, that the real GDP growth rate reached 1.9 percent (equivalent to 7.9 percent annual growth). In June 1999, Taichi Sakaiya, Minister of State for Economic Planning, declared that Japan was about coming to the end of the period of economic decline. In April 2000, Bank of Japan announced that the diffusion index (DI) for the large manufacturing sector has improved for the five consecutive quarterly periods as of January–March 2000. It appears that the need for foreign-born professionals, especially for young professionals, will increase in 2000 and thereafter, since a significant number of Japanese companies, national institutes, and universities are already internationalized and adopting global operations and management.

Many Japanese host organizations have welcomed foreign-born professionals during the past 15 years, even though many Japanese host professionals and managers were not used to dealing with them because of language barriers and cultural differences. Some foreign-born professionals and host Japanese collaborated very happily with each other, while others had less happy times because of misunderstandings and interpersonal friction resulting from cultural differences. There is no doubt, however, that there will be a growing, or at least a constant, need in the new millennium for hiring of foreign-born professionals, since Japan cannot survive without dealing with people from the rest of the world through trade, through cultural, academic, and sport activities, and through formal personnel exchange programs.

Table 1.1 shows the number of registered foreign professionals in Japan by qualification from Fiscal Year 1994 to 1998. It can be seen that, of the total of about 1.5 million foreign residents, about 41 percent and 59 percent of them are permanent residents and nonpermanent residents, respectively, and the total number of engineers and professors reached over 20,000 as of the end of 1998, while the numbers of intracompany transferees and of specialists in humanities/international services were about 7000 and 31,000, respectively, as of the end of 1998.

Table 1.2 shows numbers of registered foreign nationals by nationalities from 1996 to 1998. It can be seen that Korea and North Korea constitute over 40 percent of the total registered foreign nationals. This is

Table 1.1 Number of Registered Foreign Nationals by Qualification

Classification and Qualification	Year							
	1994	1995	1996	1997	1998			
	Number of Persons	Number of Persons	Number of Persons	Number of Persons	Number of Persons	Distribution Ratio (%)	Increase over 1997 (%)	
Total	**1,354,011**	**1,362,371**	**1,415,136**	**1,482,707**	**1,512,116**	**100**	**2.0**	
Permanent resident	**631,554**	**626,606**	**626,040**	**625,450**	**626,760**	**41.4**	**0.2**	
Nonpermanent resident	**722,457**	**735,765**	**789,096**	**857,257**	**885,356**	**58.6**	**3.3**	
(items below)								
Spouse for Japanese	231,561	244,381	258,847	274,475	264,844	17.5	-3.5	
Long-term resident	136,838	151,143	172,882	202,905	211,275	14.0	4.1	
Spouse or child of resident	53,252	56,692	60,783	64,275	65,675	4.3	2.2	
College student	61,515	60,685	59,228	58,271	59,648	4.0	2.4	
Precollege student	37,653	34,441	30,079	29,095	30,691	2.0	5.5	
Specialist in humanities* or international services	24,774	25,070	27,377	29,941	31,285	2.1	4.5	
Trainee	17,305	17,713	20,883	25,806	27,108	1.8	5.0	
Entertainment	34,819	15,967	20,103	22,185	28,871	1.9	30.1	
Engineer	10,119	9,882	11,052	12,874	15,242	1.0	18.4	
Skilled labor	6,790	7,357	8,767	9,608	10,048	0.7	4.6	
Instructor	6,752	7,155	7,514	7,769	7,941	0.5	2.2	
Spouse, etc., of permanent resident	7,002	6,778	6,460	6,325	6,219	0.4	-1.7	
Intracompany transferee	5,841	5,901	5,941	6,372	6,599	0.4	3.6	
Professor	3,757	4,149	4,573	5,086	5,374	0.4	5.7	
Others	84,479	88,451	94,607	102,270	114,536	7.6	12.0	

* Economist, sales professional, etc.

Source: Ministry of Justice, Japan, Statistics of Foreign Nationals' Residence.

Table 1.2 Number of Registered Foreign Nationals by Nationality as of the End of Each Year

Nationality	Number			Distribution Ratio (%)		
	1996	1997	1998	1996	1997	1998
Korea/North Korea	657,159	645,373	638,828	46.4	43.5	42.2
China	234,264	252,164	272,230	16.6	17.0	18.0
Brazil	201,795	233,254	222,217	14.3	15.7	14.7
Philippines	84,509	93,265	105,308	6.0	6.3	7.0
United States	44,168	43,690	42,774	3.1	3.0	2.8
Peru	37,099	40,394	41,317	2.6	2.7	2.7
Others	156,142	174,567	189,442	11.0	11.8	12.6
Total	1,415,136	1,482,707	1,512,116	100	100	100

Source: Ministry of Justice, Japan, Statistics of Foreign Nationals' Residence.

because Korea was part of Japan's territory from 1909 to the end of World War II (1945) and many Korean people moved or were forced to move to Japan during this period. After the war, many Koreans remained in Japan without changing their nationality. It is also worth noting that many of the registered foreign nationals from Brazil and Peru are people of Japanese origin.

INDUSTRY

A decade ago (1989–90), Japan's economy was booming and many Japanese companies became internationalized. Some companies such as Sony and Honda were already engaging in global operation and global management at that time. As a result, the rates and numbers of foreigners (with college or graduate degrees) hired in many industry sectors became significant, as shown in Tables 1.3 and 1.4. Those rates and numbers significantly increased during the 1990s, as many more companies became internationalized and an increasing number of companies adopted international and global management and operations to a variety of degrees, as shown in Table 1.5.

Table 1.3   Rate of Hiring of Foreigners (with College or Graduate School Degrees) by Japanese Industrial Sector during F.Y. 1989 (the Period April 1989 through March 1990)

Sector	Number of Companies Surveyed That Hired Foreigners	Percentage of Total Surveyed Companies in This Field That Hired Foreigners
Securities	22	68.8
Electronics	44	44.4
Communications	14	43.8
Commodity trading (shosha)	9	39.1
Information	19	38.8
Machinery (factory automation)	49	32.9
Machinery (shipbuilding)	12	29.3
Department stores and supermarkets	23	26.7
Automobiles	18	26.1
Banking and insurance	22	24.7
Textiles, paper, and miscellaneous consumable products	22	20.6
Metals	9	20.0
Food and pharmaceuticals	22	19.5
Chemicals	12	14.1
Transportation and travel and leisure services	15	12.7
Home construction and real estate	22	11.5
Energy	1	4.2

Note: F.Y.: Fiscal Year

Source: Nikkei Newspaper, August 3, 1990. Figures are based on hiring by companies' head offices only and do not take into account the number of foreign professionals who immediately transferred to local offices, to factories, or to an R&D center.

Figures 1.3 and 1.4, respectively, show the numbers of foreign nationals employed in 1995 and 1996 by industry sector and by size of the works. The manufacturing industry and service industry hire the largest and the second largest number of foreign nationals, respectively. Also, those works with 100 through 299 employees hired the largest numbers of foreign nationals, since

Table 1.4 Number of Foreign Employees Hired (with College or Graduate School Degrees) by Japanese Industrial Sector during F.Y. 1989 (the Period April 1989 through March 1990)

Sector	Total Number of Employees Hired	Average Number per Company
Communications	68	6.8
Electronics	150	4.8
Automobiles	73	4.3
Department stores and supermarkets	60	3.8
Commodity trading	26	3.7
Securities	70	3.7
Banking and insurance	58	3.4
Home construction and real estate	67	3.4
Textiles, paper, and miscellaneous consumable products	40	2.9
Information	36	2.8
Machinery (shipbuilding)	27	2.7
Machinery (factory automation)	94	2.5
Metals	15	2.5
Chemicals	21	1.9
Food and pharmaceuticals	25	1.8
Transportation and travel and leisure services	14	1.6
Energy	1	1.0

Source: Nikkei Newspaper, August 3, 1990. Figures are based on hiring by companies' head offices only, as in Table 1.3. The companies considered in this table are only those that gave a specific number of foreign employees in reply to the Nikkei Newspaper questionnaire. The denominator used in calculating the average number per company in each field therefore differs from that used in determining the percentage for the corresponding field shown in Table 1.3.

these works appeared to have relative difficulty finding a sufficient number of highly skilled Japanese employees.

Some large corporations hire highly skilled and/or highly promising foreign nationals at their offices in Japan and then send them back to their home countries to work as managers for the company's subsidiaries after they gain several years of experience in Japan. However, these Japanese companies might have cause for concern if those foreign nationals instead remained with the company

Table 1.5 Types and Degrees of Internationalization and Globalization Seen in Typical Japanese Companies and Industry Sectors

Type	Industry Sector	Typical Companies	Note
1. Global, overseas–oriented	Home electric appliances, computers, networks, software, program contents, automobiles, communications equipment	Sony, Honda Motor, Uniden	Advanced multinational corporation; global operation or manufacturing, sales, administration, and R&D
2. Global, based in Japan	Electrical machinery, computers, network communications equipment, cameras, office equipment, some chemicals, some automobiles, trading, financial, securities	Matsushita, Fujitsu, Hitachi, Toshiba, Canon, Kao, Dainippon Ink & Chemicals, Toyota, Mitsui & Company, Mitsubishi, Tokyo-Mitsubishi Bank, Nomura	Adopting division company systems; independent strategy for each product market in consumer goods; complying with globalization of markets, standards, trade, etc.
3. Global, capital tie-up	Automobiles	Nissan-Renault	Global alliance
4. Production in overseas market	Home appliances, consumer electronics, food, steel	Sanyo, Ezaki-Glico, Nippon Steel, Kawasaki Steel	Customer's taste–oriented; steel sheet for automobiles
5. Low-cost overseas production	Audio and communications equipment, electrical machinery, textiles & apparel, food, large-scale retail aluminum	Aiwa, Toshiba Tec, Daihen, Daiwabo, Daiei, Showa Aluminum	Percentage of overseas production depends on currency exchange rates in many cases

6. Overseas sales network	Petrochemicals	Mitsubishi Chemical	
7. Engineering network	Engineering, marine equipment	JGC (Nikki), Toyo Engineering, MODEC*	
8. R&D network	Pharmaceuticals, electronics	Takeda Chemical Industries, Fujitsu	High technology
9. Communications network	Communications	NTT, Japan Telecom	Internet and computer communication
10. Export-oriented	Shipbuilding, machine tool manufacture, robots	IHI, KHI, MES**, FANUC	Heavy and large-scale products, advanced original products, and technology, etc.
11. Domestic-based, import-oriented	Electric power and other energy distribution and sales; mining; retail; construction	Tokyo Electric Power, Kansai Electric Power, Tokyo Gas, Osaka Gas, Showa Shell Sekiyu, Tonen, Sumitomo Coal Mining, Kojima	Utilities; import of natural, energy, and mineral resources; favorable wind due to government policy for expansion of domestic demand

*MODEC: Mitsui Ocean Development Company

**IHI: Ishikawajima-Harima Heavy Industries

KHI: Kawasaki Heavy Industries

MES: Mitsui Engineering and Shipbuilding Company

Industry Sector	Abbreviation	Number of Foreign Nationals	
		Year 1996	Year 1997
Construction	Con.	2,448	2,521
Manufacturing	Man.	64,874	71,151
Transportation/Communication	Tr./Com.	4,388	4,887
Wholesale/Retail/Restaurants	W./R./R.	7,498	8,615
Service	Service	22,111	24,499
Others	Others	1,725	2,288

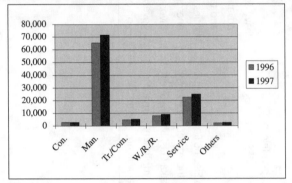

Figure 1.3   Number of foreign nationals directly employed, by industry sector, 1996 and 1997. (From Ministry of Labor, Institute of Labor Administration, *Gaikokujin Koyou Q&A* (*Questions and Answers Concerning Employment of Foreigners*), March 1998.)

in Japan for a long period of time, since it is not uncommon for foreign nationals in Japan to change jobs every one to three years, seeking better pay and benefits.

It is also a common practice for Japanese companies to hire young professionals from the host country to work for an overseas subsidiary and then send the more promising among them to Japan to educate them in the Japanese way of corporate operation and management. After a few years' experience, the Japanese companies typically send them back to the subsidiary company in the host country. In this age of central ownership combined with local decision making, this type of employment has become increasingly common and is well received by the public, but those sent to Japan from developing countries sometimes quit when they have to go back to their home countries as the working conditions there are not as good as those in

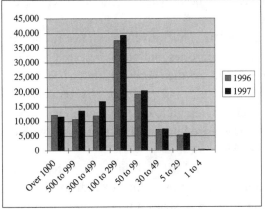

Number of Employees For Works	Total Number of Foreign Nationals Employed	
	Year 1996	Year 1997
Over 1000	11,989	11,457
500 to 999	10,480	13,414
300 to 499	11,711	16,621
100 to 299	37,364	39,190
50 to 99	19,118	20,209
30 to 49	7,084	7,201
5 to 29	5,046	5,670
1 to 4	252	199

Figure 1.4 Number of foreign nationals directly employed, by size of plant, 1996 and 1997. (From Ministry of Labor, Institute of Labor Administration, *Gaikokujin Koyou Q&A (Questions and Answers Concerning Employment of Foreigners)*, March 1998.)

Japan. Appendix 1 shows a list of companies that are hiring foreign-born professionals and those that have the capacity to hire them in the future even if they are not doing so now.

NATIONAL INSTITUTES

The Japanese Parliament unanimously passed the Science and Technology Basic Law in November 1995 to promote science and technology in the new millennium. On the basis of this law, the Japanese cabinet in July 1996 made a decision to enact the Science and Technology Basic Plan, a comprehensive action plan to promote science and technology in Japan. Under this plan, it is clearly stated that Japan must play a role of front-runner in science and technology in the global economy and thus must provide an open research community to any qualified scientists and engineers. International exchanges of scientists and research engineers are thus encouraged along with this policy, and the budgets for these activities have been increased since the Science and Technology Basic Law was passed.

Figure 1.5 shows the governmental administrative structure of science and technology in Japan. The national laboratories and public research organizations in Japan are listed in Appendix 2, and their locations, in Figure 1.6. The following describes some of the operations and activities of several of these governmental and government-related research organizations; many other agencies and organizations are characterized by similar activities but with different missions and predetermined territories.

The Science and Technology Agency and MITI (Ministry of International Trade and Industry) organizations such as the Agency of Industrial Science and Technology (AIST), among other ministries and agencies, administer the national budget and activities for a variety of national institutes and laboratories. Special organizations such as the New Energy and Industrial Technology Development Organization (NEDO) administer the research and development activities in the field of energy and industrial science and technology under the auspices of AIST and another MITI organization, the Agency of Natural Resources and Energy (ANRE). NEDO also promotes joint research between the government, industry, and academic sectors and collaborative research efforts such as those in microgravity with NASA, the German Space Agency, and the French Space Agency (CNES). The National Space Development Agency of Japan (NASDA) administers space-related research typically under the auspices of the Science and Technology Agency and also promotes joint research between NASA, the Canadian Space Agency, the European Space Agency, and the Russian Space Agency. Private sector organizations such as the Japan Space Utilization Promotion Center (JSUP) promote joint research, with a focus on microgravity and space research for industrial applications, between government, industry, and academic sectors in Japan and the rest of the world. There are, of course, some overlaps in the administrated research activities among those organizations in interdisciplinary areas.

To apply for international fellowship positions and/or to participate in international collaborative research projects, foreign-born scientists and engineers will probably have to be introduced by Japanese counterparts at national laboratories, public research organizations, and universities who well know and appreciate their professional work.

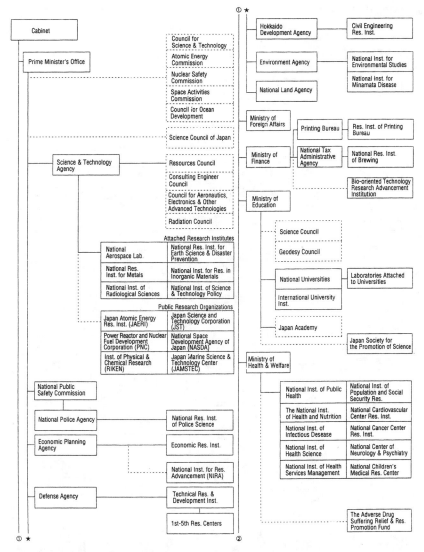

① ★

Cabinet

Prime Minister's Office

- Council for Science & Technology
- Atomic Energy Commission
- Nuclear Safety Commission
- Space Activities Commission
- Council for Ocean Development
- Science Council of Japan

Science & Technology Agency

- Resources Council
- Consulting Engineer Council
- Council for Aeronautics, Electronics & Other Advanced Technologies
- Radiation Council

Attached Research Institutes

National Aerospace Lab.	National Res. Inst. for Earth Science & Disaster Prevention
National Res. Inst. for Metals	National Inst. for Res. in Inorganic Materials
National Inst. of Radiological Sciences	National Inst. of Science & Technology Policy

Public Research Organizations

Japan Atomic Energy Res. Inst. (JAERI)	Japan Science and Technology Corporation (JST)
Power Reactor and Nuclear Fuel Development Corporation (PNC)	National Space Development Agency of Japan (NASDA)
Inst. of Physical & Chemical Research (RIKEN)	Japan Marine Science & Technology Center (JAMSTEC)

National Public Safety Commission

National Police Agency — National Res. Inst. of Police Science

Economic Planning Agency — Economic Res. Inst.

— National Inst. for Res. Advancement (NIRA)

Defense Agency — Technical Res. & Development Inst.

— 1st-5th Res. Centers

① ★   ②

- Hokkaido Development Agency — Civil Engineering Res. Inst.
- Environment Agency — National Inst. for Environmental Studies
- National Land Agency — National Inst. for Minamata Disease
- Ministry of Foreign Affairs — Printing Bureau — Res. Inst. of Printing Bureau
- Ministry of Finance — National Tax Administrative Agency — National Res. Inst. of Brewing
- Ministry of Education
  - Bio-oriented Technology Research Advancement Institution
  - Science Council
  - Geodesy Council
  - National Universities — Laboratories Attached to Universities
  - International University Inst.
  - Japan Academy — Japan Society for the Promotion of Science
- Ministry of Health & Welfare

National Inst. of Public Health	National Inst. of Population and Social Security Res.
The National Inst. of Health and Nutrition	National Cardiovascular Center Res. Inst.
National Inst. of Infectious Desease	National Cancer Center Res. Inst.
National Inst. of Health Science	National Center of Neurology & Psychiatry
National Inst. of Health Services Management	National Children's Medical Res. Center

The Adverse Drug Suffering Relief & Res. Promotion Fund

Note: The current structure is valid until the planned administrative reform becomes effective as early as 2001. (As part of the reform, national laboratories will become independent juridical persons, and the unification of Science and Technology Agency and Monbusho (shown as Ministry of Education in Figure 1.5) among others will be conducted.)

(continues)

Figure 1.5 Administrative structure of science and technology in Japan. (From Japan Science and Technology Corporation (JST), *National Laboratories and Public Research Organizations in Japan, J98S01T4000, Printed in 1998.*)

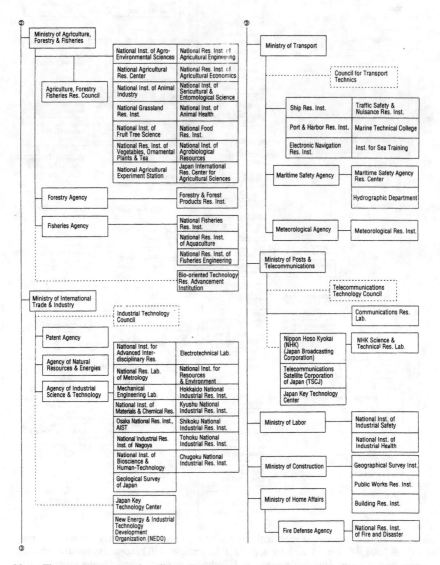

Note: The current structure is valid until the planned administrative reform becomes effective as early as 2001. (As part of the reform, national laboratories will become independent juridical persons, and the unification of Science and Technology Agency and Monbusho (shown as Ministry of Education in Figure 1.5) among others will be conducted.)

Source: National Laboratories and Public Research Organizations in Japan, Japan Science and Technology Corporation (JST), J98S01T4000, Printed in 1998

Figure 1.5   *Continued.*

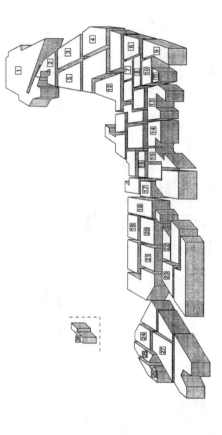

Hokkaido ☐1
Civil Engineering Research Institute, Hokkaido
    Development Bureau
Hokkaido National Agricultural Experiment Station
Hokkaido National Fisheries Research Institute
Hokkaido National Industrial Research Institute (HNIRI)
Hokkaido River Disaster Prevention Research Center

Aomori ☐2
Institute for Environmental Sciences
Aomori Advanced Industrial Technology Center

Iwate ☐3
Tohoku National Agricultural Experiment Station
Iwate Biotechnology Research Center (IBRC)
Iwate Industrial Research Institute (IIRI)

Miyagi ☐4
Tohoku National Fisheries Research Institute
Tohoku National Industrial Research Institute

Akita ☐5
Research Institute for Brain and Blood Vessels - Akita

Chiba ☐9
National Institute of Radiological Sciences (NIRS)
Central Customs Laboratory (CCL)
National Institute of Mental Health, National Center of
    Neurology and Psychiatry (NCNP)
Japan Chemical Analysis Center (JCAC)

Tokyo ☐10
National Research Institute of Police Science
National Aerospace Laboratory (NAL)
National Institute of Science and Technology Policy
    (NISTEP)
National Research Institute of Brewing (NRIB)
National Institute of Population and Social Security
    Research
National Institute of Health Services Management
    (NIHSM)
Nationa Institute of Public Health (NIPH)
Nationnl Institute of Infectious Diseases (NIID)
The National Institute of Health and Nutrition
National Institute of Neuroscience, National Center of
    Neurology and Psychiatry (NCNP)
National Cancer Center Research Institute
National Institute of Health Sciences (NIHS)

Kanagawa ☐11
Sagami Chemical Research Center (SCRC)
Kanagawa Industrial Technology Research Institute
    (KITRI)
Kanagawa Cancer Center Research Institute

Niigata ☐12
Hokuriku National Agricultural Experiment Station
Japan Sea National Fisheries Research Institute

Shizuoka ☐13
National Research Institute of Far Seas Fisheries
    (NRIFSF)

Aichi ☐14
National Industrial Research Institute of Nagoya
Japan Fine Ceramics Center (JFCC)

Mie ☐15
National Research Institute of Vegetables, Ornamental
    Plants and Tea (NIVOT)
National Research Institute of Aquaculture

Shiga ☐16
Lake Biwa Research Institute

(continues)

Figure 1.6   Locations of national laboratories and public research organizations in Japan. (From Japan Science and Technology Corporation (JST), *National Laboratories and Public Research Organizations in Japan.***)**

Ibaraki [6]
National Research Institute for Metals (NRIM)
National Research Institute for Earth Science and Disaster Prevention (NIED)
National Institute for Research in Inorganic Materials (NIRIM)
National Institute for Environmental Studies (NIES)
National Institute of Animal Industry
National Agriculture Research Center (NARC)
National Institute of Agrobiological Resources (NIAR)
National Institute of Agro-Environmental Sciences (NIAES)
National Institute of Fruit Tree Science
National Research Institute of Agricultural Engineering (NRIAE)
National Institute of Sericultural and Entomological Science (NISES)
National Institute of Animal Health (NIAH)
National Food Research Institute (NFRI)
Japan International Research Center for Agricultural Sciences (JIRCAS)
National Research Institute of Fisheries Engineering (NRIFE)
Forestry and Forest Products Research Institute (FFPRI)
National Institute for Advanced Interdisciplinary Research (NAIR)
National Research Laboratory of Metrology (NRLM)
Mechanical Engineering Laboratory (MEL)
National Institute of Materials Chemical Research (NIMC)
National Institute of Bioscience and Human-Technology (NIBH)
Geological Survey of Japan (GSJ)
Electrotechnical Laboratory (ETL)
National Institute for Resources and Environment (NIRE)
Meteorological Research Institute (MRI)
Public Works Research Institute (PWRI)
Building Research Institute (BRI)
Geographical Survey Institute (GSI)

Tochigi [7]
National Grassland Research Institute (NGRI)

Saitama [8]
The Institute of Physical and Chemical Research (RIKEN)
Japan Science and Technology Corporation (JST)
Japan Sewage Works Agency (JS)
Institute of Agricultural Machinery (IAM) / Bio-oriented Technology Research Advancement Institution (BRAIN)
Saitama Cancer Center Research Institute

Tokyo [10]
National Children's Medical Research Center
National Research Institute of Agricultural Economics
Ship Research Institute (SRI)
Electronic Navigation Research Institute (ENRI)
Traffic Safety and Nuisance Research Institute (TSNRI)
Hydrographic Department, Maritime Safety Agency
Communications Research Laboratory (CRL)
National Institute of Industrial Safety (NIIS)
National Research Institute of Fire and Disaster (NRIFD)
NHK Science & Technical Research Laboratories
Japan Atomic Energy Research Institute (JAERI)
Power Reactor and Nuclear Fuel Development Corporation (PNC)
National Space Development Agency of Japan (NASDA)
Remote Sensing Technology Center of Japan (RESTEC)
International Development Center of Japan (IDCJ)
Nippon Institute for Biological Science (NIBS)
The Research Institute of Tuberculosis / Japan Anti-Tuberculosis Association (JATA)
Japanese Foundation for Cancer Research
International Association of Traffic and Safety Sciences (IATSS)
Railway Technical Research Institute (RTRI)
Overseas Coastal Area Development Institute of Japan (OCDI)
Japan Weather Association (JWA)
Agricultural Policy Research Committee, Inc. (APRC)
Superconductivity Research Laboratory (SRL) / International Superconductivity Technology Center (ISTEC)
Institute of Research and Innovation (IRI)
Japan Automobile Research Institute (JARI)
Japan Wildlife Research Center (JWRC)
Chemicals Inspection & Testing Institute, Japan
Central Research Institute of Electric Power Industry (CRIEPI)
Tokyo Metropolitan Institute for Neuroscience
The Tokyo Metropolitan Institute of Medical Science
Tokyo Metropolitan Institute of Gerontology
The Tokyo Metropolitan Research Laboratory of Public Health

Kanagawa [11]
National Research Institute of Fisheries Science
Port and Harbour Research Institute (PHRI)
Institute for Sea Training
National Institute of Industrial Health
Japan Marine Science & Technology Center (JAMSTEC)
Japan Environmental Sanitation Center (JESC)
Kanagawa Academy of Science and Technology (KAST)

Osaka [17]
National Cardiovascular Center Research Institute
Osaka National Research Institute, AIST
Osaka Bioscience Institute (OBI)
Osaka Prefectural Institute of Public Health

Hyogo [18]
Marine Technical College (MTC)
Japan Synchrotron Radiation Research Institute

Tottori [19]
Tottori Mycological Institute (TMI)

Okayama [20]
Okayama Ceramics Research Foundation
Industrial Technology Center of Okayama Prefecture

Hiroshima [21]
Chugoku National Agricultural Experiment Station
Nansei National Fisheries Research Institute
Chugoku National Industrial Research Institute
Radiation Effects Research Foundation (RERF)

Kagawa [22]
Shikoku National Agricultural Experiment Station
Shikoku National Industrial Research Institute

Ehime [23]
Industrial Research Center of Ehime Prefecture

Fukuoka [24]
Fukuoka Industrial Technology Center (FITC)

Saga [25]
Kyushu National Industrial Research Institute

Nagasaki [26]
Seikai National Fisheries Research Institute
Technology Center of Nagasaki (TCN)

Kumamoto [27]
National Institute for Minamata Disease
Kyushu National Agricultural Experiment Station

Okinawa [28]
Akajima Marine Science Laboratory (AMSL) / Establishment of Tropical Marine Ecological Research (ETMER)

Source: National Laboratories and Public Research Organizations in Japan, Japan Science and Technology Corporation (JST), J98S01T4000, Printed in 1998

Figure 1.6 *Continued.*

UNIVERSITIES AND OTHER ACADEMIC INSTITUTIONS _____

The aforementioned Science and Technology Basic Law and Plan have also triggered budget increases in the Japanese academic sector. The Ministry of Education, Science, Sports and Culture—*Monbusho*, in Japanese—administers and promotes exchanges of scientists and research engineers with overseas academic institutions on a bilateral or multilateral basis, utilizing such international organizations as UNESCO, OECD, and APEC.

Today, Japanese professors are hiring qualified scientists and engineers from around the world under their research budgets. Hiring of foreign-born scientists and research engineers for *tenured* university faculty positions in Japan is also modestly increasing, after a few to several years' successful experience of their working at Japanese national institutes and private-sector research laboratories. Selected national universities, other public universities, and private universities that have graduate schools and/or research institutes or laboratories are shown in Appendices 3a, 3b, and 3c, respectively.

Exchange agreements of scholars between Japanese universities and overseas counterparts, and scholarships such as those provided by Fulbright programs, are also good starting points for working at Japanese universities.

Instructor positions and professorships in foreign-language studies are also open to foreign-born professionals at colleges and universities. Foreign-language teaching at secondary schools is also promoted under the Japan Exchange and Teaching (JET) Program by Monbusho with the cooperation of the Ministry of Foreign Affairs and the Ministry of Home Affairs. A total of 4831 young foreign-language instructors came to Japan from the United States, the United Kingdom, Australia, New Zealand, Canada, Ireland, South Africa, France, and Germany under this program in 1997.

The numbers of incoming foreign professors to Japan and registered foreign professors in Japan are 1309 and 4573, respectively, as of the end of 1996 and constituted 1.7 percent and 4.7 percent, respectively, of the total numbers of incoming and registered qualified professionals of that year.[2] The numbers of incoming foreign (language) instructors and registered foreign instructors in Japan are 2847 and 7514, respectively, as of the end of 1996 and constituted 3.6 percent and 8.0 percent, respectively, of the total incoming and registered instructors of that year.[3]

[2] Ministry of Justice, Japan, *Statistics of Administration of Arriving and Departing Foreign Nationals and Statistics of Foreign Nationals' Residence.*
[3] Ibid.

2

Future Trends in Employment of Foreign-Born Professionals

Hiroshi Honda

INTRODUCTION

During most of the 1990s, Japanese industry, public organizations, and universities became increasingly internationalized, and some became truly globalized, while Japan's overall economy entered into a period of slow growth. During the period from the latter half of 1997 to the end of 1998, quarterly real GDP growth rates were negative or zero, and people in the government, private, and academic sectors were struggling to regain healthy and steady economic growth. It was not until 1999 that Japan's economy appeared to have regained a positive growth, owing much to the implementation of economic reform policy by the Japanese government.

As Japan enters a period with a smaller population of younger people and a larger population of older people in this age of the global economy, the nation will increasingly need young professionals with international skills, including those from overseas countries. Figure 2.1 shows the past record and forecast figures for the labor force by age range from 1990 to 2010, substantiating the aforementioned trends.

Surveys show that Japanese organizations are typically motivated to hire foreign-born professionals for the following reasons:

1. To secure ability and knowledge that the typical Japanese lacks
2. To secure highly qualified professionals in general
3. To deal smoothly with overseas clients and organizations

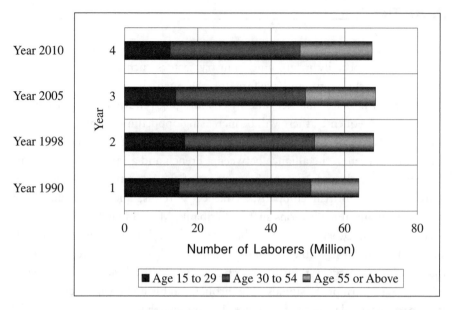

Figure 2.1   Past record and forecast for labor force by age range from 1990 to 2010. (From Koyou Shingikai (employment council), Advisory Board to Minister of Labor F.Y. 1999.)

4. To stimulate Japanese employees
5. To gain innovative and unique ideas that Japanese would unlikely have
6. To take advantage of foreign-born professionals' industriousness
7. To relocate foreign-born professionals at overseas subsidiary offices and plants
8. Because foreign-born professionals have been introduced by relevant organizations and clients
9. Requests by parent companies
10. To save on personnel costs and to achieve flexibility in number of employees during periods of recession

The aforementioned reasons have been given mostly by the industry sector. Items 1, 2, 4, and 5 would be the key and critical reasons for national institutes and universities to hire foreign-born professionals.

INDUSTRY

The Japanese government is aggressively implementing measures for revitalizing industry, especially in growing industry sectors such as information, life sciences, biotechnology, and welfare for the aged, by encouraging foreign investment in Japanese markets and by encouraging venture business by entrepreneurs. Consequently, there will be increasing opportunities for foreign-born professionals to work in Japan. Good examples of entry of foreign capital into Japan can be seen in the alliance between a French and a Japanese automobile company (Renault and Nissan); acquisition of the telecommunications company IDC by a British company, Cable & Wireless; and establishment of American financial institutions such as Citibank, Merrill Lynch, and Goldman Sachs in Japan. Other companies that are likely to hire foreign-born professionals are named in Table 1.5 and Appendix 1. It must be noted, however, that some of the large, established companies would hire (or accept to co-work with) foreign-born professionals only through established business channels. It is also of particular note, as Figure 1.4 shows, that the works with 100 through 299 employees hired the greatest number of foreign-born professionals, a majority of whom are considered to be skilled laborers, because of these plants' impending needs for qualified personnel.

In conclusion, there will be a growing or constant need for foreign-born professionals in the twenty-first century, if not the booming one seen about a decade ago, as Japan is increasingly becoming interdependent with overseas partners in this global economy.

NATIONAL INSTITUTES

The Japanese government is encouraging and promoting collaborative research and development with international partners in high-technology areas such as information science and technology, life sciences, biotechnology, energy sciences and technology, and space sciences. There will be good opportunities for working at one of the national institutes listed in Appendix 2, for qualified professionals with good interpersonal skills. Language skill is an advantage but is not mandatory.

The budget for international exchange programs is unlikely to decrease, and could increase, as indicated by Article 18 of "Chapter 4 Promotion of

International Exchange" of the Science and Technology Basic Law. According to the Science and Technology Basic Plan, the number of postdoctoral fellowships for foreign scientists and research engineers provided by the Science and Technology Agency (STA) will have increased from 340 as of 1996 to about 1000 by F.Y. 2000.

UNIVERSITIES AND OTHER ACADEMIC INSTITUTIONS _____

Since the Science and Technology Basic Law and Basic Plan were established in 1995 and 1996, respectively, university professors have come to command greater research budgets than ever before. According to the Science and Technology Basic Plan, the number of postdoctoral fellowships for foreign scientists and research engineers provided by the Japan Society for the Promotion of Science (JSPS) will have increased from 420 as of 1996 to 1050 by F.Y. 2000.

Note, however, that the search for qualified professionals for research collaboration is based mainly on qualifications. In science and technology, an emphasis is placed on whether you can conduct research successfully or not, which depends heavily on researchers' talents and skills, rather than on personnel costs.

Not all the universities and their affiliated research laboratories listed in Appendix 3 will necessarily hire scientists and research engineers; some of them may merely hire language instructors. However, the number of foreign-born professionals hired by them will likely increase in the future, either through established exchange programs or through personal introduction by professors who have a favorable assessment of a foreign-born professional's work.

UNIVERSITIES AND OTHER ACADEMIC INSTITUTIONS

Applying for a Job in Japan

3

Looking into Companies

Hiroshi Honda

GENERAL CONSIDERATIONS

Many Japanese companies will have at least a constant need for foreign-born professionals in the 2000s, even if the Japanese economy slows down at some time during this period, as described in Part 1. Especially in this age of globalization of various professions and of business practice, credentials with qualifications such as Registered Professional Engineer, Certified Public Accountant, and juris doctor will be valued at certain Japanese companies. Young Japanese people are also feeling challenged to pass the examinations for these qualifications today.

It appears that a skilled or talented foreign-born professional has a decent opportunity to find a job in Japanese industry. However, Japanese companies are also looking for candidates with personalities that they judge will fit into Japanese corporate culture and thus will be well accepted by their Japanese coworkers and managers. Personality often plays a vital role in the organization in Japanese culture. Japanese companies emphasize group effort and team spirit. In general, Japanese organizations prefer professionals with moderate (not too aggressive) personalities that can relatively easily co-work with Japanese colleagues and managers in their corporate culture. This is especially true of design, sales, and corporate administration departments; individualism is perhaps tolerated to a greater extent at laboratories and research facilities.

Another concern of personnel departments is term of employment. Through their experiences in dealing with foreign-born professionals for the past decade, they came to hire those professionals on a contractual basis with a limited period such as half a year, one year, or the duration of a project. They will, of course, renew the contracts of foreign-born professionals who show competency in job skills and performance, as long as the Japanese company has work and projects that require foreign-born professionals' skills.

Though the job hopping prevalent overseas is well known in Japan and has also become common in Japan, especially among the younger generation, Japanese employers by and large still hope that any qualified professionals, once they become well accepted in their workplace, will plan for a fairly long-term or a lifelong commitment—at least three years, if not more.

In this chapter, I will focus on how one can find a desirable position in Japanese industry. Note, however, that a position offered by a company may not initially seem attractive to a foreign-born professional. Even so, the employee will have a chance to move to another position or to work on another job if he or she shows competency at his originally assigned job. The candidate must bear in mind that the company's motivation, just like that of companies at home, is to maximize output; a Japanese company, in hiring foreigners, has the added desire to achieve a positive influence on its Japanese employees. Therefore, the candidate's career plan must be compatible with the company's hiring policy.

EFFECTIVE METHODS OF APPLICATION[1]

A foreign-born professional seeking a job in Japan with a Japanese company should know the preferred method of hiring in Japan. Seldom would an applicant's letter or application sent directly to a company in Japan be successful in establishing the necessary dialogue leading to employment. Japanese society relies relatively heavily on introductions from known associates before a "new face" is invited to join an existing group or activity; this rule applies as much to obtaining any professional position in a Japanese company as to any other activity.

---

[1] The author would like to note that portions of the section Effective Methods of Application are based on a discussion with Mr. Raymond C. Vonderau, who coauthored Chapter 3 for the first edition of this book (1992).

There are, of course, a number of hiring, recruiting, and head-hunting companies that will list their advertisements for job opportunities for foreign-born professionals in popular journals and newspapers available in Japan. This also substantiates that job-hopping is also becoming popular in Japan.

The surest first step would be to obtain employment with a Japanese affiliate company located in one's home country. Experience and a good job performance record with a Japanese affiliate company are important assets, and employment with the affiliate provides an opportunity for the home-country company managers to become acquainted with the applicant's abilities and career objectives. They can then make introductions and recommendations to specific managers in the company's Japan offices.

For a recent graduate, the college or university placement office can often provide a list of Japanese companies that have affiliates located in the applicant's home country. It is particularly helpful if a professor who is familiar with the applicant's academic record has previously recommended candidates who were subsequently hired by the affiliate company. A professor who has done consulting work for a company in Japan or the affiliate of the Japanese company would also be in a position to help. Again, the personal contact and recommendation are most helpful in ensuring consideration for employment.

Some professionals have also become acquainted with Japanese companies by attending trade fairs organized for the purpose of interviewing seniors or recent graduates having a specific kind of experience and an interest in working in a foreign country. This method of meeting qualified professionals is often used by Japanese companies that have no affiliate company outside of Japan.

If none of these routes is open, one should write for advice either directly to companies' personnel departments in Japan or to an alumnus of one's school working in Japan. Here, expert knowledge or experience will be an advantage. Since Japan is facing a shortage of young professionals, especially in high-tech fields and in business administration, expertise in these will make an applicant very attractive. Knowledge of Japanese will also be an asset. Some (established) companies, however, strictly apply the same hiring criteria for candidates of all nationalities, such as hiring of fresh graduates from graduate and undergraduate schools only and those with high-level competency in the Japanese language.

Besides obtaining a professional position with a Japanese company, the professional, making the commitment to move to Japan to work, needs to understand, at least in general terms, the following important matters relating to job satisfaction and performance discussed throughout this book:

1. Job level and promotions
2. Compensation practices
3. Importance of supporting the group effort

COMPENSATION AND PROMOTION

Tables 3.1a, 3.1b, 3.2, and 3.3 are good references for foreign-born professionals regarding representative salaries. Japanese companies had encouraged the custom of lifetime employment through a system of awarding step-by-step raises on the basis of seniority. However, this tradition began to lose its hold in the 1990s, especially for middle-aged and senior employees, because of the slow economy during this period in Japan. It can be seen from Tables 3.1a and 3.1b that the seniority-based salary increase rates for employees aged 45 through 55 for given industry sectors are lower in F.Y. 1998 than those in F.Y. 1990 in general. However, younger-generation employees secure pay increases even though an increasing amount of job hopping exists among young Japanese and among foreign-born professionals—a result of the globalization of the job market and industry.

Salary usually decreases after age 55 if a person is not a *yakuin* (a member of the board of directors, including managing director, vice president, president, and chairman). The model salary is the before-tax income. In the case of foreign professionals, the local tax is applied only to those who have lived in Japan for one year or longer. Japan has concluded tax treaties with a number of nations; therefore, foreign professionals may be exempted from their home country's taxes, or they may need to pay to the home country the difference between the Japanese taxes and their home country's corresponding taxes. The model salary does not include the payment for the overtime work that Japanese professionals regularly do at the direction of their supervisors. If the payment for overtime is included, the actual salary will range from 110 to 130 percent of the model salary, depending on the company. However, overtime payment has been tight, especially during the last half of the 1990s, at most Japanese companies. The annual salary may reach almost twice as much as the

Table 3.1a Average Model Annual Salary for Manufacturing Industries (College Graduated)

(F.Y. 1990: for Companies with 1000 or More Employees; F.Y. 1998: for All Companies Surveyed)
Currency Conversion Rate: $1 (U.S.) = 135 yen for F.Y. 1990; $1 (U.S.) = 120 yen for F.Y. 1998

Industry Sector	Fiscal Year	Currency	Age 22	Age 25	Age 27	Age 30	Age 35	Age 40	Age 45	Age 50	Age 55
All industries (average)	F.Y. 1998	$U.S.	24,958	31,946	—	42,989	53,038	64,926	76,880	87,594	94,928
		1000 yen	2,995	3,834	—	5,159	6,365	7,791	9,226	10,511	11,391
Manufacturing (average)	F.Y. 1990	$U.S.	16,267	24,704	28,881	33,281	42,511	52,644	64,230	77,289	84,630
		1000 yen	2,196	3,335	3,899	4,493	5,739	7,107	8,671	10,434	11,425
	F.Y. 1998	$U.S.	25,038	31,978	—	43,061	52,207	64,193	77,315	88,040	95,843
		1000 yen	3,005	3,837	—	5,167	6,265	7,703	9,278	10,565	11,501
Chemicals	F.Y. 1990	$U.S.	16,600	26,289	31,422	36,970	47,970	58,193	69,467	85,319	91,763
		1000 yen	2,241	3,549	4,242	4,991	6,476	7,856	9,378	11,518	12,388
	F.Y. 1998	$U.S.	25,267	32,317	—	45,128	54,268	67,330	80,416	91,124	96,649
		1000 yen	3,032	3,878	—	5,415	6,512	8,080	9,650	10,935	11,598
Electrical machinery	F.Y. 1990	$U.S.	16,667	25,437	28,815	33,022	40,815	49,044	54,526	67,259	—
		1000 yen	2,250	3,434	3,890	4,458	5,510	6,621	7,361	9,080	—
	F.Y. 1998	$U.S.	24,397	30,962	—	42,365	51,091	58,644	71,897	81,020	83,564
		1000 yen	2,928	3,715	—	5,084	6,131	7,037	8,628	9,722	10,028
Food and fishery	F.Y. 1990	$U.S.	15,993	25,000	30,000	34,311	44,119	57,089	71,689	84,200	89,526
		1000 yen	2,159	3,375	4,050	4,632	5,956	7,707	9,678	11,367	12,086
	F.Y. 1998	$U.S.	25,376	32,328	—	43,520	53,720	71,200	85,082	97,406	109,823
		1000 yen	3,045	3,879	—	5,222	6,446	8,544	10,210	11,689	13,179
Glass and ceramics	F.Y. 1990	$U.S.	16,185	23,748	27,133	31,296	40,044	50,822	64,400	75,519	91,541
		1000 yen	2,185	3,206	3,663	4,225	5,406	6,861	8,694	10,195	12,358
	F.Y. 1998	$U.S.	23,965	30,721	—	39,368	41,363	48,484	60,845	70,345	—
		1000 yen	2,876	3,687	—	4,724	4,964	5,818	7,301	8,441	—
Machinery	F.Y. 1990	$U.S.	16,133	23,467	26,815	30,348	36,770	48,919	61,452	68,133	77,970
		1000 yen	2,178	3,168	3,620	4,097	4,964	6,604	8,296	9,198	10,526
	F.Y. 1998	$U.S.	24,813	31,594	—	42,126	51,774	63,258	73,453	84,432	89,988
		1000 yen	2,978	3,791	—	5,055	6,213	7,591	8,814	10,132	10,799

(continues)

Table 3.1a (continued)

Industry Sector	Fiscal Year	Currency	Age 22	Age 25	Age 27	Age 30	Age 35	Age 40	Age 45	Age 50	Age 55
Nonferrous metals and	F.Y. 1998	$U.S.	25,638	32,450	—	41,021	51,734	74,393	89,811	99,533	108,089
metal products		1000 yen	3,077	3,894	—	4,923	6,208	8,927	10,777	11,944	12,971
Oil products	F.Y. 1998	$U.S.	26,134	34,108	—	47,188	61,991	76,675	85,800	90,300	98,250
		1000 yen	3,136	4,093	—	5,663	7,439	9,201	10,296	10,836	11,790
Precision instruments	F.Y. 1990	$U.S.	16,837	25,681	31,385	36,356	47,059	54,889	81,711	94,504	91,556
		1000 yen	2,273	3,467	4,237	4,908	6,353	7,410	11,031	12,758	12,360
	F.Y. 1998	$U.S.	25,493	33,550	—	45,586	54,408	63,188	72,293	87,843	101,645
		1000 yen	3,059	4,026	—	5,470	6,529	7,583	8,675	10,541	12,197
Pulp and paper	F.Y. 1998	$U.S.	25,185	32,368	—	38,959	43,881	50,941	56,915	69,416	78,827
		1000 yen	3,022	3,884	—	4,675	5,266	6,113	6,830	8,330	9,459
Steel	F.Y. 1990	$U.S.	16,274	25,074	28,919	34,311	43,933	53,570	57,637	81,037	
		1000 yen	2,197	3,385	3,904	4,632	5,931	7,232	7,781	10,940	
	F.Y. 1998	$U.S.	24,853	31,973	—	43,660	49,760	56,487	65,995	76,000	84,760
		1000 yen	2,982	3,837	—	5,239	5,971	6,778	7,919	9,120	10,171
Textile and apparel	F.Y. 1990	$U.S.	16,148	24,896	29,696	33,926	43,281	55,415	68,044	77,667	78,259
		1000 yen	2,180	3,361	4,009	4,580	5,843	7,481	9,186	10,485	10,565
	F.Y. 1998	$U.S.	24,890	32,177	—	43,634	51,282	58,078	74,780	86,978	92,530
		1000 yen	2,987	3,861	—	5,236	6,154	6,969	8,974	10,437	11,104
Transportation equipment	F.Y. 1990	$U.S.	16,126	23,822	26,763	30,141	37,481	46,578	53,067	69,015	70,481
		1000 yen	2,177	3,216	3,613	4,069	5,060	6,288	7,164	9,317	9,515
	F.Y. 1998	$U.S.	24,720	31,369	—	40,942	50,719	56,989	71,673	80,958	85,953
		1000 yen	2,966	3,764	—	4,913	6,086	6,839	8,601	9,715	10,314
Other manufacturing	F.Y. 1998	$U.S.	26,538	33,987	—	43,043	55,643	64,362	76,870	88,788	96,543
		1000 yen	3,185	4,078	—	5,165	6,677	7,723	9,224	10,655	11,585

Average model annual salary = standard monthly salary (F.Y. 1990 or F.Y. 1998) × 12 + winter bonus (1989 or 1998) + summer bonus (1990 or 1998).

Source: Seisansei Model, Sougou Chingin Jittai Chosa, Nihon Seisansei Honbu (Productivity Model, Total Salary Investigation, by Japan Productivity Center) for F.Y. 1990 data.
Nenkan Chingin Shoyo no Jittai '99 Edition, Romu Ghosei Kenkyusho (Annual Salary and Bonus, by Institute of Labor Administration) for F.Y. 1998 data
Note: F.Y. Starts April 1 and ends on March 31 in the following year.

Table 3.1b Average Model Annual Salary for Non-Manufacturing Industries (College Graduated)

(F.Y. 1990: for Companies with 1000 or More Employees; F.Y. 1998: for All Companies Surveyed)

Currency Conversion Rate: $1 (U.S.) = 135 yen for F.Y. 1990; $1 (U.S.) = 120 yen for F.Y. 1998

Industry Sector	Fiscal Year	Currency	Age 22	Age 25	Age 27	Age 30	Age 35	Age 40	Age 45	Age 50	Age 55
Nonmanufacturing (average)	F.Y. 1998	$U.S.	26,538	33,987	—	43,043	55,643	64,362	76,870	88,788	96,543
		1000 yen	3,185	4,078	—	5,165	6,677	7,723	9,224	10,655	11,585
Construction	F.Y. 1990	$U.S.	15,881	25,356	29,481	33,919	42,822	52,926	63,311	74,141	83,096
		1000 yen	2,144	3,423	3,980	4,579	5,781	7,145	8,547	10,009	11,218
	F.Y. 1998	$U.S.	23,048	29,596	—	38,682	49,779	57,582	67,656	76,470	83,178
		1000 yen	2,766	3,552	—	4,642	5,974	6,910	8,119	9,176	9,981
Finance and insurance	F.Y. 1998	$U.S.	21,165	31,085	—	44,436	59,434	75,923	81,324	92,856	91,855
		1000 yen	2,540	3,730	—	5,332	7,132	9,111	9,759	11,143	11,023
Land transport	F.Y. 1998	$U.S.	24,761	30,013	—	40,405	48,112	40,955	57,133	66,283	70,195
		1000 yen	2,971	3,602	—	4,849	5,773	4,915	6,856	7,954	8,423
Marine and air transport	F.Y. 1998	$U.S.	26,506	33,007	—	42,330	59,337	77,668	90,863	103,973	109,282
		1000 yen	3,181	3,961	—	5,080	7,120	9,320	10,904	12,477	13,114
Mass communication	F.Y. 1998	$U.S.	32,831	42,992	—	59,408	73,468	86,628	94,492	98,578	97,855
		1000 yen	3,940	5,159	—	7,129	8,816	10,395	11,339	11,829	11,743
Retail	F.Y. 1998	$U.S.	24,179	30,942	—	42,369	54,362	64,963	75,415	88,239	97,908
		1000 yen	2,902	3,713	—	5,084	6,523	7,796	9,050	10,589	11,749
Services	F.Y. 1998	$U.S.	26,070	33,484	—	44,253	56,913	69,823	80,426	90,112	100,374
		1000 yen	3,128	4,018	—	5,310	6,830	8,379	9,651	10,813	12,045
Warehouse and distribution	F.Y. 1998	$U.S.	26,551	33,885	—	46,351	60,633	73,958	85,018	97,183	95,653
		1000 yen	3,186	4,066	—	5,562	7,276	8,875	10,202	11,662	11,478

Average model annual salary = standard monthly salary (F.Y. 1990 or F.Y. 1998) × 12 + winter bonus (1989 or 1998) + summer bonus (1990 or 1998).

Source: Seisansei Model, Sougou Chingin Jittai Chosa, Nihon Seisansei Honbu (Productivity Model, Total Salary Investigation, by Japan Productivity Center) for F.Y. 1990 data.
Nenkan Chingin Shoyo no Jittai '99 Edition, Romu Ghosei Kenkyusho (Annual Salary and Bonus, by Institute of Labor Administration) for F.Y. 1998 data.

Table 3.2 Average Model Annual Salary for Managers in the Industry Sector (College Graduates) Currency Conversion
Rate: $1 (U.S.) = 135 yen for F.Y. 1990; $1 (U.S.) = 120 yen for F.Y. 1998

| Position | | Currency | Number of Employees per Company | | | | | | |
| English | Japanese | | Over 5000 | Over 3000 | 1000–4999 | 1000–2999 | 300–999 | Few than 1000 | Fewer than 300 |
			F.Y. 1990	F.Y. 1998	F.Y. 1990	F.Y. 1998	F.Y. 1990	F.Y. 1998	F.Y. 1990
Department general manager	Bucho	$U.S.	90,593	108,867	76,148	98,663	65,556	83,213	58,519
		1000 yen	12,230	13,064	10,280	11,840	8,850	9,986	7,900
Department deputy general manager	Bu-jicho	$U.S.	—	94,714	—	85,836	—	72,396	—
		1000 yen	—	11,366	—	10,300	—	8,687	—
Section manager	Kacho	$U.S.	68,074	81,700	60,000	72,895	50,519	64,629	44,519
		1000 yen	9,190	9,804	8,100	8,747	6,820	7,756	6,010
Section deputy manager	Ka-jicho	$U.S.	—	74,029	—	67,091	—	56,585	—
		1000 yen	—	8,884	—	8,051	—	6,790	—
Group leader	Kakaricho	$U.S.	53,556	54,964	44,741	54,681	38,296	46,914	34,000
		1000 yen	7,230	6,596	6,040	6,562	5,170	5,630	4,590

Average model annual salary (F.Y. 1990 or F.Y. 1998) = standard monthly salary (F.Y. 1990 or F.Y. 1998) × 12 + winter bonus (1989 or 1998) + summer bonus (1990 or 1998).

Note: Salaries for department deputy general manager and section deputy manager are obtained by multiplying those of general manager and the coefficients given in the source.

Source: Seisansei Model, Sougou Chingin Jittai Chosa '91 Edition, Nihon Seisansei Honbu (Productivity Model, Total Salary Investigation, by Japan Productivity Center) for F.Y. 1990 data. Nenkan Chingin Shoyo no Jittai '99 Edition, Romu Ghosei Kenkyusho (Annual Salary and Bonus, by Institute of Labor Administration) for F.Y. 1998 data.

Table 3.3 Percentage Index of Average Model Annual Salary by Age of Employee and Size of Company (College Graduates)
Percentage Index of 100 for the Average Salary of Companies with Employees of Over 3000 for F.Y. 1998
Percentage Index of 100 for the Average Salary of Companies with Employees of Over 1000 for F.Y. 1990

| | Number of Employees per Company | | | | | | |
| | Over 3000 | Over 1000 | 1000–2999 | 300–999 | Few than 1000 | 100–299 | Fewer than 100 |
Age	F.Y. 1998	F.Y. 1990	F.Y. 1998	F.Y. 1990	F.Y. 1998	F.Y. 1990	F.Y. 1990
22	100	100	101	97	98	95	93
25	100	100	98	94	92	94	93
27	100	100	—	91	—	90	88
30	100	100	96	90	86	89	89
35	100	100	95	87	84	86	86
40	100	100	92	86	79	84	83
45	100	100	88	86	76	82	78
50	100	100	92	85	78	82	78
55	100	100	90	86	77	83	74

Average model annual salary (F.Y. 1990 or F.Y. 1998) = standard monthly salary (F.Y. 1990 or F.Y. 1998) × 12 + winter bonus (1989 or 1998) + summer bonus (1990 or 1998)

Source: Seisansei Model, Sougou Chingin Jittai Chosa '91 Edition, Nihon Seisansei Honbu (Productivity Model, Total Salary Investigation, by Japan Productivity Center) for F.Y. 1990 data. Nenkan Chingin Shoyo no Jittai '99 Edition, Romu Ghosei Kenkyusho (Annual Salary and Bonus, by Institute of Labor Administration) for F.Y. 1998 data.

model salary if the employee is in an extremely fast-growing industry sector such as communications-related software, just as in the United States.

Table 3.2 shows average model salaries of department general managers *(bucho)*, department deputy general managers *(bu-jicho)*, section managers *(kacho)*, section deputy managers *(ka-jicho)*, and group leaders *(kakaricho)* of companies of different sizes. In the case of bucho, bu-jicho, and kacho, who would not be members of a union, there is generally no compensation for overtime. Table 3.3 shows average model salaries according to age for different-sized companies. As the tables show, the salaries paid by larger companies are usually higher than those found in smaller companies.

Tables 3.4 and 3.5 show the average monthly salaries and the average annual salaries for academic positions, such as dean *(gakubucho)*, professor

Table 3.4 Average Monthly Salary for Academic Positions in Japan, F.Y. 1998

Position	Japanese Term	Currency	Age									
			20–24	24–28	28–32	32–36	36–40	40–44	44–48	48–52	52–56	56–
Dean	Gakubucho	$U.S.	—	—	—	—	—	—	5,740	6,697	6,760	6,736
		1000Yen	—	—	—	—	—	—	689	804	811	808
Professor	Kyoju	$U.S.	—	—	—	—	4,732	5,081	5,372	5,768	6,001	6,182
		1000Yen	—	—	—	—	568	610	645	692	720	742
Associate professor	Jokyoju	$U.S.	—	—	3,553	3,849	4,246	4,547	4,774	4,989	5,239	5,388
		1000Yen	—	—	426	462	510	546	573	599	629	647
Lecturer	Koushi	$U.S.	—	2,425	3,140	3,431	3,806	4,229	4,491	4,789	4,855	5,008
		1000Yen	—	291	377	412	457	508	539	575	583	601
Research associate	Joshu	$U.S.	1,763	2,357	2,868	3,185	3,647	3,955	4,309	4,542	5,258	4,275
		1000Yen	212	283	344	382	438	475	517	545	631	513
High school principal	Kouchou	$U.S.	—	—	—	—	—	—	5,847	5,586	6,286	6,057
		1000Yen	—	—	—	—	—	—	702	670	754	727
High school vice principal	Kyoutou	$U.S.	—	—	—	—	—	—	4,554	5,150	5,482	5,502
		1000Yen	—	—	—	—	—	—	547	618	658	660
High school teacher	Kyoushi	$U.S.	2,204	2,382	2,739	3,217	3,656	4,100	4,418	4,783	4,929	5,135
		1000Yen	265	286	329	386	439	492	530	574	592	616

Source: Romugyoseikenkyusho (The Institute of Labor Administration), Shokushubetsu Kyuyo Jittai '99 Edition, p. 40.

Table 3.5 Average Annual Salary (Including Two Bonuses) for Academic Positions in Japan

Position	Japanese Term	Currency	Age									
			20–24	24–28	28–32	32–36	36–40	40–44	44–48	48–52	52–56	56–
Dean	Gakubucho	$U.S.	—	—	—	—	—	—	99,123	115,656	116,738	116,334
		1000 yen	—	—	—	—	—	—	11,895	13,879	14,009	13,960
Professor	Kyoju	$U.S.	—	—	—	—	81,729	87,751	92,775	99,620	103,633	106,771
		1000 yen	—	—	—	—	9,807	10,530	11,133	11,954	12,436	12,813
Associate professor	Jokyoju	$U.S.	—	—	61,355	66,467	73,336	78,526	82,443	86,161	90,473	93,051
		1000 yen	—	—	7,363	7,976	8,800	9,423	9,893	10,339	10,857	11,166
Lecturer	Koushi	$U.S.	—	41,872	54,229	59,251	65,722	73,042	77,564	82,704	83,841	86,484
		1000 yen	—	5,025	6,507	7,110	7,887	8,765	9,308	9,924	10,061	10,378
Research associate	Joshu	$U.S.	30,448	40,701	49,538	55,001	62,977	68,300	74,415	78,439	90,801	73,837
		1000 yen	3,654	4,884	5,945	6,600	7,557	8,196	8,930	9,413	10,896	8,860
High school principal	Kochou	$U.S.	—	—	—	—	—	—	100,983	96,468	108,564	104,608
		1000 yen	—	—	—	—	—	—	12,118	11,576	13,028	12,553
High school vice principal	Kyotou	$U.S.	—	—	—	—	—	—	78,654	88,935	94,671	95,027
		1000 yen	—	—	—	—	—	9,439	10,672	11,361	11,403	—
High school teacher	Kyoushi	$U.S.	38,069	41,137	47,298	55,552	63,144	70,810	76,305	82,608	85,129	88,687
		1000 yen	4,568	4,936	5,676	6,666	7,577	8,497	9,157	9,913	10,216	10,642

Average annual salary = average monthly salary (F.Y. 1998) ¥ (12 + 5.27 (average rate for sum of winter and summer bonuses for F.Y. 1997)).

Source: Romugyoseikenkyusho (The Institute of Labor Administration), Shokushubetsu Kyuyo Jittai '99 Edition, pp. 40, 77.

(kyoju), associate professor *(jokyoju)*, lecturer *(koushi)*, research associate *(joshu)*, high school principal *(kocho)*, high school vice principal *(kyotou)*, and high school teacher *(kyoushi)* in Japan. In F.Y. 1989, salaries for academic positions were considered to be lower than those for equivalent, industry-based positions. However, as can be seen from Tables 3.1 through 3.5, average annual salaries for senior university-based positions were higher than those in industry by F.Y. 1998. This was due to slow economic growth and subsequent low pay increases or even pay decreases in industry from the mid- to late 1990s in Japan.

It must be noted that the conversion ratio of the yen to the U.S. dollar fluctuated significantly for the past decade, as shown in Figure 3.1. Therefore, the salaries in U.S. dollars for F.Y. 1990 (from April 1990 to March 1991) shown in Tables 3.1 through 3.5 are based on a rate of 135 yen per dollar, while the currency conversion rate went down to below 80 yen per dollar in the first half of 1995 and then went up to 145 yen per dollar in August 1998. However, the rate from October 1998 to July 1999 was kept around 120 yen per dollar, owing much to the policy set by the Ministry of Finance and Bank of Japan. Thus, this rate is used for computing the salaries in U.S. dollars for F.Y. 1998 (from April 1998 to March 1999), shown in Tables 3.1 through 3.5. As of April 2000, the Japanese yen has become stronger as the Japanese economy

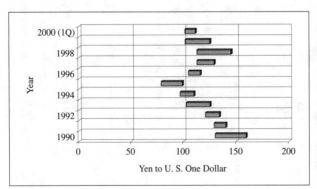

Year	Average
2000 (1Q)	105–107
1999	115.3*
1998	130.9
1997	121
1996	108.8
1995	94.1
1994	102.2
1993	111.2
1992	126.7
1991	134.7
1990	144.8

(*July–August Source: IMF)

Source: World Economic Outlook, IMF (October 1999) and White Paper on World Economy, Economic Planning Agency, Japan (F. Y. 1998) for Annual Average Conversion Rates Except for that of 2000

Figure 3.1 Variation of conversion rate of Japanese yen to the U.S. dollar, 1990–99.

appears to be regaining its strength relative to its U.S. counterpart toward the turn of the century.

The salaries of foreign professionals are usually set on a short-term basis (two or three years) or on a project basis. A foreign-professional's salary can therefore be higher than the salary of a Japanese employee having equal qualifications. Japanese companies are willing to bring into their regular employment systems any foreign-born professionals who are well accepted among their Japanese colleagues. However, these people usually prefer employment on a temporary basis, with its higher salary levels; therefore, the salary level for the foreign-born professional varies extensively with the employee's qualifications.

In contrast, however, Japanese college graduates of the same age are generally treated equally at the initial stage of employment regardless of whether they have bachelor's, master's, or doctor's degrees. Any extra years after high school spent in preparation for college entrance examinations are not taken into consideration, even though about half of high school graduates fail when they first apply for admission to universities. This is because the competency and skills of the individuals depend also on many other factors, such as the environment in which the employee was brought up and potential capabilities that cannot be measured with academic degrees. It is generally advantageous to hold advanced degrees for positions at research laboratories; for administration and design departments, those with master's or bachelor's degrees would be preferred; and for factory positions, those with bachelor's or master's degrees would be preferred. In many Japanese manufacturing companies, those with bachelor's degree only from factory positions are commonly elected to be the company president. Advancement thereafter depends solely on the individual's performance and the company's policy over a long period of employment.

It is not uncommon today for qualified foreign professionals to occupy middle to senior management positions at industry and professorial positions at universities in Japan. It is rare, however, to see foreign professionals occupy the top management positions at Japanese industry, unless sent by overseas companies that have a capital alliance with a Japanese counterpart. Such examples can be seen in President Mark Fields of Mazda, which is in a capital alliance with Ford Motor Company in the United States, and President and Chief Operating Officer Carlos Ghosn of Nissan Motors, which is in a capital alliance with Renault, France. Employees at Japanese leading companies would have a high mental resistance against seeing foreign nationals (or even seeing Japanese nationals from outside their own companies) as their presidents, as is the case in other countries to various degrees.

Fringe benefits provided by Japanese firms usually include coverage of commuting expenses; the use of inexpensive company dormitories, with breakfast and supper provided, for single persons; inexpensive (sometimes free, at local sites) sports and leisure facilities; and use of resort facilities at discount rates. Reasonably priced health insurance and paid holidays are also offered. It must be noted, however, that the content of the fringe benefit package varies substantially among companies. Foreign professionals should therefore ask the personnel department about a company's fringe benefit system.

4

Finding a Job at a National Institute

Hiroshi Honda

GENERAL CONSIDERATIONS

Many national institutes have been accepting an increasing number of foreign-born professionals for their fellowship positions, since the Science and Technology Basic Law was passed in 1995 and the Science and Technology Basic Plan was enacted in 1996, as stated in Chapter 2. The duration of stay in Japan for postdoctoral fellowships is typically one or two years with a possible extension, while that for an invitation fellowship ranges from two weeks to nine months. The purpose of these fellowships is to promote international collaboration in basic areas of research conducted at the national institutes, introduced in Chapter 1 and listed in Appendix 2, and in academic institutions.

It does not appear to be common for a foreign-born professional (even if that person is a Japanese national who has worked in industry or academia in Japan) to occupy a permanent position at a national institute, probably because national security, in either a broad or a narrow sense, may be affected by the institutes' activities. However, those who have successfully and fruitfully conducted research at these institutes will afterward have a good opportunity to work in Japanese industry or universities.

Important Areas of Research Identified by the Science and Technology Agency

A survey conducted by the Science and Technology Agency (STA) has identified the following as important areas for research and development[1]

---

[1] *STA Today*, February 1999.

1. Preventing destruction of the global environment
2. Conquering heretofore incurable diseases
3. Constructing disaster-resistant cities
4. Birth of new industries
5. Creating a comfortable environment for senior citizens
6. Development of inexpensive, good-quality products
7. An affluent society form cultural perspectives

STA also promotes international cooperation on research programs in the following areas:

1. Science of the human frontier
2. International space station
3. International thermonuclear experimental reactor
4. Scientific and technological cooperation and support for the Asia-Pacific region, the former Soviet Union, and Central and Eastern European countries

Areas of Research Administered by the Agency of Industrial Science and Technology, MITI

The Agency of Industrial Science and Technology (AIST) of MITI identifies the following areas of past and ongoing research as important for creating new industries and revitalizing existing industries:[2]

1. Superconductivity materials and devices
2. Materials science: research in such areas as high-performance materials for severe environments, nonlinear photonics materials, advanced chemical processing technology, silicon-based polymers, synergy ceramics, and technology for novel highly-functional materials
3. Biotechnology: marine biotechnology (producing fine chemicals from marine organisms and tropical bioresources), molecular assemblies for functional protein systems, production and utilization technologies of complex carbohydrates, and evolutionary molecular engineering

[2] *Industrial Science and Technology Frontier Program (ISTF), AIST, 1996.*

4. Electronics: information and communication technology such as new models for software architecture, quantum functional devices, ultimate manipulation of atoms and molecules, femtosecond technology, and human media technology

5. Machinery and aerospace technology: super- or hypersonic propulsion systems, micromachine technology

6. Natural resources development technology, such as the manganese nodule mining system

7. Technology for human, life, and society, such as development of underground space development technology and human sensory measurement application technology

8. Medical, health care, and welfare technology: microsampling and microanalysis techniques for blood and other tissues; high-performance 3-D imaging systems for medical diagnosis; support systems for less invasive surgery; artificial organs; technical aids for self-sufficiency (user-friendly self-care equipment); technical aids for home care; multimedia systems for the handicapped; esophageal vocalization systems; comprehensive systems for supporting wheelchairs; fundamental research on medical welfare equipment technologies; collection, analysis, and distribution of information on medical welfare equipment; promoting the development of practical medical welfare equipment; and research into improving home care equipment systems for effective use of energy

AIST also focuses on the following as leading research *(sendo kenkyu)* areas:

1. Metal revolution (leaching technology of optimum composition of solvent by sunshine radiation, bacteria's oxidizing effect on metals, and investigation on state of metal ions in mixed solvent systems, selectivity in metal crystallization, etc.)

2. Advanced metrological technology

3. Harmonized molecular materials (fabrication of new materials, whose size and/or phase are controlled by molecular interactions between organic, inorganic/metalic, and polymer substances. Aimed at development of new organic/inorganic hybrid materials and realization of biomimetic functions)

4. Supermetal

5. Brainware

6. Cell handling technology
7. Photon technology
8. Smart structural systems
9. Computer chemistry for material design
10. Frontier carbon technology
11. Technology for decoding and utilization of genome information
12. Hard electronics – (fabrication technology for high quality crystal substances, device processes, characterization technology for material properties, device structures and processes)
13. Friendly network robotics
14. Technology for measuring and evaluating human behavior and cognition

Distribution of Disciplines among Scholars Invited by the Japan Society for the Promotion of Science

The Japan Society for the Promotion of Science (JSPS), a Monbusho-affiliated organization, also promotes academic research in many frontier areas and encourages international collaboration, inviting scholars from around the world. The following shows a percentage distribution of the scholars invited to universities and university-affiliated research organizations in Japan, by research field, as of April 1, 1998:[3]

1. Mathematical, physical, and engineering sciences, 41.9 percent
2. Biological, agricultural, and medical sciences, 23.7 percent
3. Humanities and social sciences, 18.8 percent
4. Chemical sciences, 15.6 percent

Other ministries and agencies, shown in Figure 1.5, also offer fellowship programs.

Number of Fellowships Offered by STA, AIST, and JSPS

The following shows the numbers of fellowship visitors by agency and ministry:

---

[3] JSPS home page, August 1999.

1. 428 researchers from 50 countries in F.Y. 1997 for STA fellowship programs
2. About 50 postdoctoral researchers per year for AIST fellowship programs, 40 researchers for industry technology fellowship programs in F.Y. 1995, and invitations extended to 30 administrators and 100 researchers (in countries receiving Official Development Assistance (ODA) to AIST laboratories for Institute for Transfer of Industrial Technology (ITIT) fellowship programs
3. 280 postdoctoral fellowship visitors (and additional invitation fellowship visitors) in F.Y. 1999 for JSPS programs

EFFECTIVE METHODS OF APPLICATION

Administrative Organizations

The aforementioned fellowship programs are administered by the following organizations:

1. The Japan Science and Technology Corporation (JST) for STA fellowships
2. New Energy and Industrial Technology Development Organization (NEDO) for MITI fellowships such as the aforementioned AIST fellowships
3. Japan Society for the Promotion of Science (JSPS) for fellowships with Monbusho sponsorship

For fellowships financially supported by other agencies and ministries, the applicants should contact those organizations shown in Figure 1.5 and Appendix 2.

Multilateral Programs

There are two types of promotion of international collaboration in research—bilateral collaboration and multilateral collaboration—through international organizations such as the United Nations, the Organization for Economic Cooperation and Development (OECD), and the Asia-Pacific Economic Cooperation (APEC). Each ministry or agency has its own policy in promoting international collaboration and cooperation, and applicants should contact the organizations of their choosing for details.

Bilateral Programs

In the case of bilateral collaboration, there are always counterpart organizations such as the National Science Foundation (NSF), the National Institute for Standards and Technology (NIST), and the Social Science Research Council (SSRC) in the United States; the Centre National de la Recherche Scientifique (CNRS) in France; Deutsche Forschungsgemeinschaft (DFG) in Germany; the Royal Society (RS)/British Council (RC) in the United Kingdom; Korea Science and Engineering Foundation (KOSEF); Department of Science and Technology of India (DST); the National Natural Science Foundation of China (NSFC); and the Japan-European Research Cooperative Program for countries excluding France, Germany, and the United Kingdom. Applicants should contact these organizations for details.

Instruction and Effective Methods of Application

Japanese governmental and government-related organizations have instruction materials for foreign-born professionals to apply for fellowship positions, which are distributed to university professors and scientists and research engineers at national institutes in Japan and their counterpart organizations in overseas countries.

For instructions, applicants can also access the Web sites of, among others, STA, MITI, and Monbusho—respectively,

http://www.sta.go.jp/

with keyword *fellowship,*

http://www.aist.go.jp/www-e/aistfelp.html

and

http://www.jsps.go.jp/e-fellow/main.html

It is usually recommended that foreign-born professionals find a Japanese partner in applying for a fellowship position or for acceptance in international cooperation-oriented projects at Japanese national institutes and/or universities. Their applications are more likely to go smoothly and to succeed if Japanese scientists and/or research engineers know their professional work well, have an interest in collaborating with them, and are willing to help them in applying for the positions.

COMPENSATION

Compensation will depend on the fellowship awardees' qualifications and their host organizations. JSPS's compensation for an incoming two-year post-doctoral fellow for F.Y. 2000 (April 1, 2000, through March 31, 2001) is given here for illustration:

1. A round-trip air ticket for the fellow only
2. A monthly maintenance stipend of 270,000 yen (U.S. $570 at a rate of 105 yen/dollar)
3. A settling-in allowance of 200,000 yen
4. Domestic research travel allowance of 58,500 yen per year
5. A monthly housing subsidy not to exceed 100,000 yen
6. A monthly family allowance of 50,000 yen if accompanied by dependents
7. Accident and sickness insurance coverage for the fellow only

Note: For fellows nominated by the European Commission, the related expenses other than a monthly stipend of 270,000 shall be borne by the E.C.

In addition, the fellow can also apply for Monbusho's Grant-in-Aid for JSPS Fellows (Tokubetsu Kenkyuin Shorei-hi) of a maximum of 1,500,000 yen through his or her host organization.

For invitation fellowship programs for research, JSPS covers the following expenses for the fellow:

1. For short-term (14 to 60 days) invitation fellowships (195 vacancies for F.Y. 1998), a round-trip air ticket, per diem of 18,000 yen, and domestic travel allowance
2. For long-term (6 to 10 months) invitation fellowships (58 vacancies for F.Y. 1998), a round-trip air ticket, monthly stipend of 270,000 to 300,000 yen, housing allowance, and domestic travel allowance

The compensation will vary, as stated above, but should be sufficient to cover the necessary expenses.

5

Finding a Job at a Japanese University

Shuichi Fukuda

In the Meiji era, that is, in the late nineteenth century, many of the teachers at national universities came from western countries. Similarly, far back in history, in the sixth and early seventeenth centuries, much of what was important in Japanese culture as well as much of Japanese technology was brought to Japan by the Chinese and Koreans, who played the roles of teachers at that time.

In modern times, however, it was very difficult until quite recently for a foreigner to find a position in a Japanese university, even though many of the universities were established only after World War II. There have long been foreign-born professors teaching in Japan, but in most cases only at missionary or private universities. In contrast to the Meiji era, national and municipal universities have closed their doors to non-Japanese; those few fortunate enough to find positions were in most cases hired as assistants; it was almost impossible to become a full faculty member.

A new law enacted in 1982 has gone into effect that stipulates that foreigners may be employed in national and municipal universities under the same conditions of employment as Japanese. This law provides that a foreigner may obtain the position of professor, associate professor, or lecturer and may participate in faculty meetings with all privileges, such as voting, just as Japanese may. The term of employment is left to the individual university. In most universities, the term for a foreign faculty member is two or three years, after which the contractee has to reapply. But wages are increased in subsequent terms.

The unusual and difficult experiences foreigners are likely to have working in Japanese university laboratories are well described in the article "Maximizing Mutual Benefits for Your Stay in a Japanese Laboratory," by Professor Robert G. Latorre, in the book *Science in Japan*.[1] The article and the book provide much valuable information on how to apply for research jobs in universities, national institutes, and private industry. Those who are interested in obtaining a research position in Japan should give the book at least a cursory glance before applying.

If you wish to obtain a research position in a Japanese university, the most important thing you have to do is to find a professor who shares your interest in your research plans and who is willing to take charge of your application. This is true even for Japanese who are applying for university positions, even though positions are advertised publicly and are theoretically open to all applicants. In Japan, everything is run on a person-to-person basis, so that it is vital to find a professor interested in employing you or willing to introduce you to friends and colleagues.

But you have to be very careful to take into account the Japanese way of management in a research setting, as you will understand after reading *Science in Japan*. In Japan, professors (and researchers) in principle, do not move from one university to another. Most of them work at the university from which they have graduated; therefore, you cannot simply file an application one day and begin work the next. If you really wish to work in Japan, you must remember that not only you but also the employer will have to make preperations. This is the main difference between universities in the western world and those in Japan. Every decision is made on the assumption of a lifelong (or at least long-term) commitment. Thus, if the university you wish to work for is not interested in your research subject, you will have to either persuade people there to take interest in it or find another theme. In the former case, you will have to make a commitment to stay for a long time; if you then plan to move to another university for a short stay, you will have to be prudent because of the potential problem. It is customary for a Japanese professor to take the initiative in finding a replacement.

These principles are fundamentally the same for national, municipal, and private universities. But private universities are subject to far less strict laws

---

[1] Robert S. Cutler (ed.), *Science in Japan: Japanese Laboratories open to U.S. Researchers,* Technology Transfer Society, Indianapolis, 1989.

regarding the employment of foreigners. They can hire anyone they can afford if they judge him or her worthy. The same flexibility applies for Japanese applicants, too. Private universities are also more flexible as far as research environment is concerned. But private universities have their disadvantages, too. In most cases there are more students in private universities, so that more time has to be spent on teaching than in research. In national and municipal universities, annual budgets are determined by law and very difficult to modify; thus, the number of professors and students are limited. Therefore, generally you will have more time for research at national universities. Yet this budgetary situation in the public universities limits flexibility in employment of foreigners and the selection of research subjects. You cannot make a research proposal in Japan as freely as you can in a western country. Researchers who share laboratories in Japan share equipment, assistants, advice, and even glory to some extent. Thus, you will have to convince the other workers in your laboratory of the worth of your project, or else you will have to modify your objectives, settling on a less ambitious theme or one that fits more closely into the existing operations of the laboratory.

I cannot overemphasize how essential it is to make sure that the laboratory situation at your chosen university is compatible with your objectives, because once you have begun working at one university laboratory in Japan, you are likely to find being admitted to another a prohibitively time-consuming task requiring delicate manuevering. If you come to Japan with the express purpose of working at more than one laboratory, you should make this clear at the very beginning. Your advisers will then try to find positions which best fit your situation. In all probability these would be lecturer's or professorial posts, where you would have the rights guaranteed under the 1982 law and a salary typical of that paid to a foreign born faculty members. In such a post, however, you will be allowed more flexible management of your laboratory and, therefore, closer adherence to your research interests. But such positions are few and hard to get.

To summarize, it is very important that you maintain an attitude of cooperation both when you apply for a position and afterward when you begin working. Even if the position you are offered seems to be a little different from the one you want, take the opportunity and work within the existing framework at the new laboratory. This is the Japanese approach to research; it will allow your colleagues to feel most comfortable with you. Of course, you should be firm about your opinions, but you should not stick with them to the last. You should listen to all objections, assess the situation, and try to solve the problems together with the other researchers in the laboratory. And remember that in Japan, you are expected to create the situation you want where you are, rather than look for it somewhere else.

Facets of
Japanese
Society

6

Overview of Japanese Society in the Global and Asian Communities[1]

Hiroshi Honda

COMPARATIVE FEATURES OF JAPAN

Japan is a small island country adjacent to the Asian continent with a history extending back about 2000 years, as shown in Appendix 4, Concise History of Japan. The ancestors of the Japanese included aborigines on the islands and people who later came from Korea, mainland China, Mongolia, and Southeast Asia, among other places. The Japanese people were greatly influenced by China and Korea until about 1000 A.D., or the mid-Heian period. For example, Japan's written language, in Chinese characters, is gramatically similar to Korean language, and its religions of Confucianism and Buddhism were transmitted from Korea, China, and India on the Asian mainland.

Japan has certain characteristics in common with Great Britain. Both the British and the Japanese have a reputation for attaching great importance to common sense; also, Great Britain, like Japan, is a small island country adjacent to a continent. In addition to the aforementioned Asian countries, Japan has been influenced by Europe since guns were introduced by the Portuguese in 1543, and by the United States since Commodore Perry visited Japan with warships in 1853; and the country has assimilated much from cultures all over the world. Similarly, Great Britain has been strongly influenced by the European continent and has also borrowed from many other cultures. In the

---

[1] Please refer to Appendix 4, the history table, and the maps included in Appendices 7 and 8 when reading this chapter.

process, the Japanese adopted many customs, ideas, and beliefs from other countries. For example, the Japanese enjoy French, Italian, Chinese, American, German, Korean, and Indian food, among others, in addition to Japanese food. Buddhism, Shintoism, and some sects of Christianity are common in Japan, and Japanese people tend to mix different religious ceremonies. For example, a majority of Japanese families conduct the marriage ceremony in the Shinto style, (no Buddhist style), and conduct their memorial ceremony in Buddhist style, no Shinto. Some Japanese families conduct their marriage ceremony and their memorial ceremony in Christian style. These practices might appear odd to westerners, but Japanese people tend to adopt favorite customs from different options. In the view of an eminent Buddhist in Japan, all religions strive for the same goal but through different approaches and styles, and the Japanese perhaps unconsciously recognize the common ground when they practice their mixture of ceremonies.

Japan's greatest economic interdependence is with the United States and with other Asia-Pacific economies. Japanese people enjoy a friendship with the French on the cultural, scientific, technological, and interpersonal levels. Japan was much influenced by Germany technologically and scientifically, especially before World War II.

In spite of their eclectic tastes, the Japanese are obviously distinct in nature from the people of other nations. It is generally said in Japan that the Japanese do not appear to be aggressive because they tend to assert themselves indirectly and modestly in everyday life. The Japanese see themselves as rather restrained and indirect, and Americans as unrestrained and direct, when the two peoples' social standards are compared. It is generally accepted that Japan is a homogeneous, conservative society descended from an agricultural people, while most western nations are made up of descendants of hunting people.

HISTORICAL BACKGROUND: FROM RECENT CENTURIES TO MODERN TIMES

In the Edo period (from 1603 to 1868), the Tokugawa feudal government adopted a policy of spreading Confucianism in Japanese society in order to maintain its power as long as possible. (Edo is the former name of Tokyo, where the governmental administration was located, away from the imperial palace in Kyoto.) This philosophy of the ideal of a peaceful and just society still holds sway in Japan, as it does over much of Asia. Under the principles of Confucianism, elderly people are always to be respected and children are to obey parents' directions and opinions. Family responsibility is also important under Confucianism. If a member of a family commits a crime, for example, all family members share in the penalty to some degree. Because of this philosophy, family ties became very strong in Japan.

The samurai occupied the highest social rank. A samurai was obliged to serve only one sovereign and devote his life to him. Changing his allegiance meant betrayal of his society and was punished by banishment from society.

The farmer was next to the samurai in social rank. In Japanese agrarian life in particular, strong cooperation among farmers was an absolute necessity in case they needed to cope with adverse climate and weather. In times of calamity, farmers had to share the burden, and so they developed a solid team spirit. No single person, no matter how brilliant or capable, could be better off than his peers. The farmers had to share their wealth as well as their burdens. The *nanushi*, or village headman, was succeeded by his direct heir generation after generation, regardless of the heir's capability. If the family had no son, then often a capable adopted son or a son-in-law succeeded the nanushi. *Gonin-gumi* (five-person teams) were formed so that every family's behavior could be observed by representatives of four other families, thus maintaining the social order. Because of this system, families had to take special care not to wrong their neighbors or even to stand out from the community, since this could lead to jealousy and envy.

An individual's shame was a family's shame and often a local community' shame. When a community was disgraced by the conduct of the samurai, he had to restore its honor by committing harakiri to compensate for the shame. On the other hand, ordinary people in such cases were punished by authorities in public. (This traditional Japanese ethic was illustrated in a scandal, after it was revealed in 1987 that a subsidiary of Toshiba Corporation had violated the rules of COCOM, the Coordinating Committee on Export Controls. The chairman and the president of Toshiba resigned, presumably as a result of political pressure, even though the home office had not been involved in the affair. However, they were allowed to assume presidency of public service corporations such as the Electronics Industries Association of Japan, or EIAJ.

Individuals or families that wished to initiate a new enterprise could only do so with the consent of the rest of the five-person team or of the community; in addition, one family's failure was usually considered the team's failure. Denunciation and criticism were taboo, since both denouncing and denounced persons were considered disreputable. A common proverb in Japan, "Silence is golden, talkativeness evil," probably stemmed from this attitude.

Human relations were thus very (and often too) close and caring, and therefore confined—certainly by American standards—as a result of the social system and ethics of the Edo period and perhaps also because of the high population density in Japan. (On the positive side, the restrictive standards minimized the crime rate and maintained the social order.) The groundwork of present-day Japanese society was laid during this period—the "vertical" structure of Japanese society, as described by Chie Nakane (See Chapter 7 of this book).[2]

Tokyo: A Capital of Internationalized Japan

Also, during the Edo period, international trade was prohibited, to prevent the influence of outside social systems and religions, such as Christianity, from taking hold. But in the Meiji era (from 1868 to 1912), which began after the Tokugawa feudal government was overthrown, western culture streamed into Japanese society in pace with the increase in international trade. It was at the beginning of the Meiji era that Edo was renamed Tokyo, which literally means "east capital," and became capital of Japan. Western democracy and liberalism as well as western goods gradually penetrated into Japanese society. However, the Japanese emperor was still regarded as a god by all Japanese. Japan went through two industrial revolutions, like the one that had occurred earlier in the west, one in the late 1800s another in the early 1900s, though it was not until 1945, the end of World War II, that democracy took root in Japanese society. It was the military occupation under Douglas MacArthur that approved the Japanese constitution and the postwar political system and reduced the status of the Japanese emperor to just a symbol of the nation. It was only recently that some political parties seriously began to consider amending the constitution.

In the years since 1980 in particular, Tokyo has become a westernized and internationalized city with distinctive features as Japan has become the second largest economic power in the world. Changes in working life have taken

[2] Chie Nakane, *Tate-shakai no Ningen Kankei* ("human relations in the vertical society"), Kodansha, Tokyo, 1967.

place, too. However, the atmosphere and manners of the established core society are still largely reminiscent of those established in the Edo period, especially in smaller cities and rural areas.

With the internationalization of Tokyo, the Tokyo International Forum was built in the Marunouchi area, heart of Tokyo, and with the waterfront development plan, new infrastructure was constructed in the submetropolitan bay area to mediate the dense population and business activities in the Tokyo metropolitan area.

Cities Surrounding the Tokyo Metropolitan Area

Yokohama City, Kanagawa Prefecture, is a capital of the prefecture and a seaport, which deals in the largest quantity of container cargos and deals in the third largest quantity of imported, and exported and domestic goods in the nation, after No. 1 Chiba and No. 2 Nagoya Seaports. In addition to the commercial activity, an exotic atmosphere such as that found in residential areas inhabited by westerners or in Chinatown attracts many visitors and tourists. Minato Mirai 21, or Future Seaport 21, is a project under way to expand modern infrastructure for business in Yokohama.

Chiba City is also famous as a conservative city and the capital of Chiba Prefecture, with the newly built Makuhari business area, which has various facilities such as Makuhari Messe (exhibition hall) and Chiba Marine Stadium. Tokyo Disneyland is located in Urayasu City, and between there and Chiba City is Yatsu Tideland in Narashino City, where wild birds visit from Australia and other countries overseas. Choshi City is a mecca for soy sauce production. All of these cities are located in Chiba Prefecture.

Urawa is the capital of Saitama Prefecture, and Mito and Tsukuba are the capital and a research and university town, respectively, of Ibaraki Prefecture. Gunma Prefecture has produced a number of Japanese prime ministers such as Yasuhiro Nakasone, Takeo Fukuda, and Keizo Obuchi, and also is famous for wives ruling over their husbands. Tochigi Prefecture is blessed with natural hot spring, ski, and skate areas.

Other Important Centers of Japanese Culture

Kyoto literally means the "capital of capitals" (or "metropolis and capital") and was the capital of Japan for 1100 years until the end of the Edo period.

People in Kyoto take special pride in their ancient tradition. They jokingly claim that to be qualified as a "real" Kyotoite one must be descended from families living in Kyoto since before Ohnin-no-ran (a rebellion that occurred between 1467 and 1477 during the Ohnin era). Even to be considered a "real" Tokyoite or Edokko (in Japanese), one's family must have lived in Tokyo for three generations or longer. Kyotoites are known for being gentle in manner but cool toward newcomers at first. However, after a substantial period of acquaintance ranging from several years to some decades, a new person can gain acceptance as a member of this traditional society.

Osaka, located only 30 kilometers from Kyoto, is a city of merchants with an open atmosphere. People still vividly remember a housewife in Osaka who gave President Bill Clinton, during his visit to Japan, quite a verbal shot regarding his recent scandal, saying, "How did you apologize to your wife? I would not have forgiven you if I were Mrs. Clinton." Very close to Osaka is Kobe, a famous seaport with an exotic, western atmosphere. Nara was the capital of Japan from 710 to 784, before Kyoto. In the Kyoto-Osaka-Kobe-Nara (Kansai) region are the remains of ancient capitals from times before the Nara period. This region is especially noted for mainstream Japanese culture and traditions and has thus attracted many visitors from overseas.

Many other cities that can be found on the maps in Appendix 7 are also noteworthy as centers of Japanese culture and traditions. Kanazawa, Ishikawa Prefecture, for example, is a famous castle town with a conservative atmosphere on the side of Honshu toward the Sea of Japan; the area was blessed with very abundant rice production and was governed by Lord Maeda in the Edo period. People here are noted for being very polite and humble and for treating guests with special hospitality. Kenrokuen in Kanazawa is regarded as one of the three finest Japanese gardens. Fukuoka is a center of culture, commerce, and transportation on Kyushu and also a castle town that was governed by Lord Kuroda in the Edo period. The culture of Fukuoka as well as other cities of northern Kyushu has been much influenced since ancient times by the continental Asian culture, as can be easily understood from their geographical location. Sendai is a center of culture, industry, and administration in the Tohoku region, governed by Lord Date in or before the Edo period. Sapporo City is a center of commerce and political administration of Hokkaido; it became modernized through development during and after the Meiji era.

JAPAN AND THE ASIAN COMMUNITY

As Japan became increasingly interdependent with other Asian and Australesian economies, interactions with these economies, especially with the People's Republic of China, Taiwan, and Korea, accelerated. These nationalities are distinct from each other, even though they have cultural and other characteristics in common. In religion, for example, they have Buddhism and Confucianism in common, and they all originated as agricultural peoples, as opposed to hunting peoples, and share a rice-based diet.

Figures 6.1, 6.2, and 6.3 show the results of a survey on perceptions that Japanese, Chinese, and Koreans, respectively, have of Japanese, Chinese, Koreans, and Taiwanese, reported by Bu Yan.[3] Japanese are commonly viewed by neighboring Asians as industrious, modest, effort-making, and positive about working, but as neither generous nor sociable nor friendly nor bold. Chinese are viewed as broad-minded, acceptant, capability-oriented, generous, friendly, kind, and bold, but as neither modest nor positive in their attitude toward work. Koreans are viewed as industrious, fond of study, and effort-making, but are viewed as neither broad-minded nor modest. Taiwanese are viewed as fond of study and friendly, but neither modest nor generous nor kind. These characteristics are, in general, true of the respective nationalities (but of course, not of each individual) and stem from such factors as historical background or geographical conditions as they affected those nations over a long period of time.

Apart from the aforementioned characteristics, each nation has its own set of delicate feelings regarding the other nations, similar to the traditional rivalries between French, Germans, and British and to the anti-American strain among Canadians and Mexicans. The most delicate and profound feelings would be those held by older people in the People's Republic of China and Korea stemming from Japan's activities during the Pacific War (as World War II is known in the west). The Taiwanese are rather sympathetic to the Japanese and do not appear to hold much antagonism toward them, in contrast to the other two nations in this regard. However, over the more than half century

---

[3] Reprinted from Bu Yan, Author for "East Asian" of Chap. 4 Features of Foreign Employees from Various Regions of the World in H. Honda (ed.), *Honne de Ikasu Gaikokujin Shain* ("successfully dealing with foreign employees with honne), Nikkan Kogyo Shimbun (Book Division), 1994.

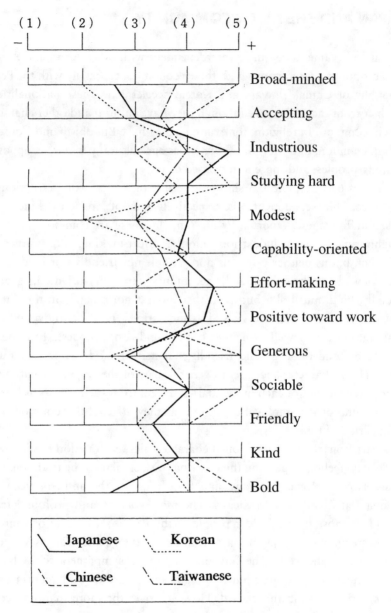

Figure 6.1   Perception of East Asians held by Japanese (survey by Bu
Yan). [From H. Honda (ed.), *Honne de Ikasu Gaikokujin
Shain* ("successfully dealing with foreign employees with
honne"), Nikkan Kogyo Shimbun, 1994.]

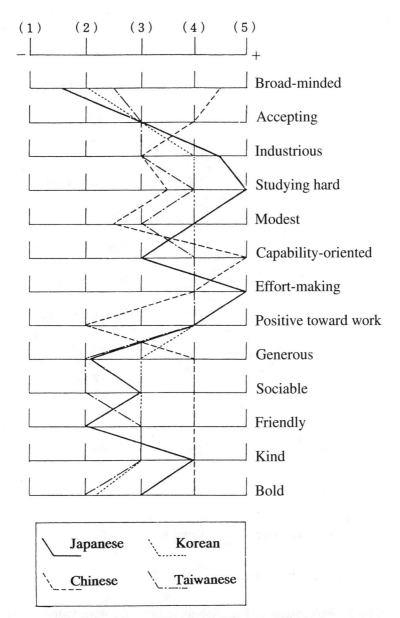

Figure 6.2 Perception of East Asians held by Chinese (survey by Bu Yan). [From H. Honda (ed.), *Honne de Ikasu Gaikokujin Shain* ("successfully dealing with foreign employees with honne"), Nikkan Kogyo Shimbun, 1994.]

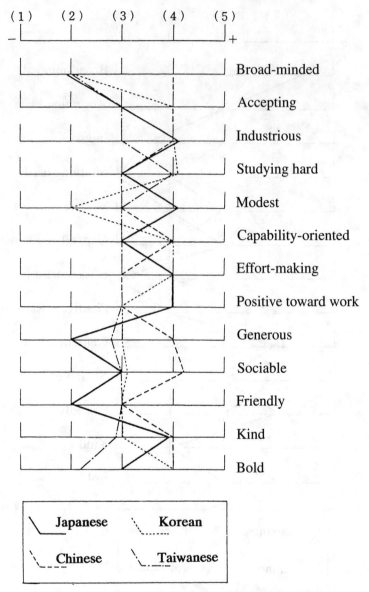

Figure 6.3   Perception of East Asians held by Koreans (survey by Bu
Yan). [From H. Honda (ed.), *Honne de Ikasu Gaikokujin
Shain* ("successfully dealing with foreign employees with
honne"), Nikkan Kogyo Shimbun, 1994.]

since the war, official development assistance (ODA), economic cooperation, and joint ventures have smoothed Japan's relationship with Korea and mainland China. The Chinese are conscious and proud of having been the center of their world for thousands of years, and many Japanese keep this national awareness in mind in dealing with Chinese people. Other East Asian businesspeople as well try to stay aware of the national pride of the people they are dealing with.

THE MODERN WORK ENVIRONMENT IN JAPAN _____

The Japanese have a saying that "capable hawks hide their talons," the meaning of which should be obvious from previous parts of this chapter. A Japanese tends to behave in a way that is inoffensive and often complimentary to other people, not only to bosses but also to junior or senior colleagues. A polite Japanese would say, "I agree with you in a sense; however, there is another idea, like this…," even if he or she actually rejects your idea. A Japanese seeks to gain the support of other people and hopes it will lead to substantial cooperation from a majority of the members of the organization and, in turn, promotion. In contrast, promotion in an organization in western society tends to rely more on an individual boss's preference.

A member of the Japanese elites is expected to make considerable sacrifice, or at least appear to the surrounding people to be doing so. Such a person performs many tasks that come up for little monetary compensation but is rewarded with the satisfaction of having the respect of senior and junior colleagues. Respected department managers are willing to sweep the floor ahead of their subordinates, when it is time for the department members to clean the office (as they indeed do in Japan).

There is no specific job description for each individual worker, and in most cases an individual's responsibilities are not clearly defined; however, the mission of a department or office is clearly stated. Therefore, the person who contributes the most to a department or office will be promoted first, but with only a modest pay increase. No one, no matter how professionally capable, will be promoted ahead of his or her peers without having both their professional recognition and respect. Harmony is valued, not competition, within a group. "He is working hard, has a nice personality, and is also well accepted by his peers" is the kind of evaluation that is often heard of a man who attains the greatest success. This means that the brightest are not always the ones who are promoted first. (Under these circumstances, it is also true that less capable persons tend to think that they have as much chance to be promoted as any-body, and indeed, much to their relief, they often are promoted early in their careers along with the best in their department.)

Nemawashi (literally, "root binding," or gaining a consensus of the neces-sary people) is also an important process in Japanese society. Nemawashi takes much time and effort. However, once it is completed, it can result in very quick action, or it can mean that a project or policy will be kept alive even if prob-lems come up of the kind that in the west would usually result in cancellation. *Nominucation* (a slang combination of the Japanese word *nomi*, "drinking," and the English word *communication*) is a very important part of nemawashi. What is said during nominucation rarely surfaces in the more formal atmos-phere of the office, but nemawashi would not function well without it. People who are good at nemawashi tend to be promoted fast in Japanese society.

This has been the traditional work environment and still holds among senior employees and managers in their forties and fifties, who were brought up under the system of company loyalty and lifetime employment during the period of high economic growth until the 1980s. As Japanese society went through a long economic slump from late 1991 to the begin-ning of 1999, this work environment has appeared as a bottleneck for vig-orous economic growth.

Job-hopping became common in the 1980s and 1990s, particularly among the younger generation, who have become less interested in maintaining the strong human relations that were a traditional force in the workplace and instead seek new opportunities to develop their talents and interests outside the organization. It was reported in 1999 that over 30 percent of young Japan-ese people who entered companies in the latter half of the 1990s quit their jobs within a few years to find new opportunities, instead of continuing to serve as a gear within a large corporate machine. These people are of a generation born and brought up after Japanese society had become rich, and thus tend to feel less attraction to a work environment that does not fit their interests or tastes, no matter how stable or stately the organization may be. Stories are told about a group of young bankers who started a gyoza (Chinese baked ground meat and vegetables wrapped in a wheat crepe) restaurant, and of someone who left a job as an inspector for a large construction company to enter a school for traditional woodworking, hoping to become a professional artist. Then, as the Japanese economy deteriorated during a few periods in the 1990s and many senior people in their late forties to early fifties were either persuaded or pressured to leave their jobs, the strong ties within Japanese companies loosened further.

When playing on a softball team at Pennsylvania State University, I noted that the batting order announced for the next game was constantly being changed by the team captain according to the performance of each player in the last game. Players who fell to low positions in the batting order seemed to take their responsibility less seriously and just enjoyed playing the game. It seemed to me that much of the way American society works fits a similar pattern. In contrast, Japanese society tended to maintain established "batting orders" until it became truly necessary to change them. However, this tendency appears to be giving way toward the American system of promotion, especially among the quickly growing companies in the high-technology sector such as communication, information, and software technology companies. Many giant companies such as Nippon Steel Corporation, Mitsubishi Heavy Industries, and Hitachi Limited are restructuring and reducing their labor force and encouraging and sometimes forcing their employees in their late forties to early fifties to leave, in order to cut costs. Similarly, global competition has forced many years ago the western giants such as IBM and General Electric to restructure their organizations and let their employees go to the companies that have bought their divisions.

FOREIGNERS IN JAPANESE SOCIETY

After reading the preceding sections, some westerners will feel that they will be unable to conform to Japanese society. However, non-Japanese-speaking people are treated as guests for their first years in Japan, especially at universities and national institutes, and many of them have an enjoyable life in Japan. When the newcomers first arrive, their Japanese hosts observe how these people fit in; subsequently, their degree of acceptance will depend in various degrees on the work environment and the individuals' personality and capability. The foreigners most successful in Japan, of course, are those who have the ability to communicate in Japanese, in addition to their expertise, capability, or knowledge. Once a foreigner can communicate in Japanese, he or she has an invaluable tool that will draw respect from the Japanese people, thus making it easy to be accepted. Good examples are Musashimaru and Akebono, grand champion, or yokozuna, sumo wrestlers from Hawaii; Kent Gilbert, a television personality and lawyer from Utah; and university president Gregory Clark from the United Kingdom. There are also numerous western businesspeople who have established their own companies after resigning from major securities corporations in the west who can speak fluent Japanese and who understand Japanese culture.

The Japanese are, in many cases, emotional rather than rational in making a decision. Once the Japanese feel that a foreigner truly understands and appreciates the Japanese mind and customs, they will respect that person and open their minds regardless of his or her nationality. This is just as true in other countries. The Japanese government welcomes the contribution of outstanding foreign nationals, once it is judged that they will be personally well accepted by Japanese people. Very good examples can be seen in American Ambassadors Edwin Reischauer and Mike Mansfield, and Sadaharu Oh, a Chinese national who was a world record holder in hitting home runs and is currently manager of a professional baseball team Daiei Hawks—among others who gained national recognition in Japan.

7

Historical Sketch and Culture of Japanese Industry

Kazuo Takaiwa and Hiroshi Honda

STAGES IN INDUSTRIAL DEVELOPMENT SINCE WORLD WAR II

It was only after the end of the Second World War in 1945 that Japanese industries began to walk alone. Before and during the war, they had faithfully delivered gigantic, ultramodern battleships, such as the Musahai and Yamato, and some of the fastest and highest-performance fighter planes, such as the Zero and Shidenkai. However, industry was the slave of the military establishment; its independence came in 1945.

The postwar years can be divided into the following general eras.

1945–1950

Immediately after the war, the Japanese people concentrated their efforts on rehabilitating their devastated land and cities with aid from the Government Account for Relief in Occupied Areas (GARIOA) fund administered by the United States. Individual survival was a struggle during these grim years.

The ravaged earth produced little, and food was constantly in short supply. Meanwhile, laborers, newly awakened to their civil rights, began to organize into unions. Labor disputes broke out often. However, in keeping with the Japanese principle of *wa* (harmony), unions and companies usually cooperated in finding solutions to the disputes. Workers did not consider it a threat when their leaders were hired into management. It was an era when production was maintained only by human power, because machine power was scarce.

1950–1955

War broke out between North and South Korea in 1950. The Japanese islands became a logistic base for the United Nations forces. At the same time, Japan was becoming ready to wake up, to stand again and walk by itself.

Japan was faced with a gap in its technical and industrial knowledge and had to dip into its still depleted purse for funds to fill the gap with advanced technology from American and European companies, without regard for the cost. Japan has little natural resources, and the leaders of the nation understood very well that the country could not presume upon the GARIOA fund forever. Thus, Japan entered an era of technology transfer and promotion of employment. Industries invested in worker training and gradually modernized their factories, and in the process many of the small-scale factories that had dominated Japanese production since the beginning of industrialization during the Meiji era were phased out.

1955–1965

Coastal industrial zones were rapidly developed, with the construction of steel mills, oil refineries, and other petrochemical complexes. Shipyards were enlarged, eventually to scales enabling the construction of mammoth tankers. Gigantic blast furnaces were purchased and installed in more and more steel mills. Hydraulic and steam power plants were established. Automobile manufacturing swelled. Unemployment, an endemic problem since the war, went through a steady fall. Thus began the era of high growth. Industries needed foreign engineers in those heady days, but only as short-term advisers.

1965–1970

In the next phase the domestic market for industrial products and general production capacity became saturated, resulting in heightened competition among different manufacturers, engineering firms, and construction companies. Contractors began increasingly to look abroad for projects, and exports in general expanded. Japanese sales forces emphasized their strong points of rapid delivery, high quality, manufacturers' willingness to adapt to local standards, and use of electronic data processing to improve service. This era marked the beginning of sharp competition among Japanese makers, design of products for foreign consumers, and cultural and trade friction. Industries began to employ foreign engineers under long-term conditions.

1970–1975

A steadily increasing number of atomic plants went into operation, while pollution and other negative effects of the high level of development began to loom as a public issue. A recession occurred after the leap in oil prices. Antipollution technologies were developed, alongside a general, ongoing automation, computerization, and streamlining of industry and construction, pushed by changes in the yen-dollar exchange rate, from fixed rate of 360 yen = U.S. $1 to step-by-step yen's appreciation, and the steadily growing price of energy.

1975–1985

Companies began to export large-scale production facilities in module form on contract for overseas buyers, and eventually Japanese companies began to shift production abroad. The emphasis in manufacturing moved to zero defects, zero accidents, and total quality control, thanks largely to the total productive maintenance (TPM) program. This was an era of large-scale production and improvements in quality. The number of foreign engineers increased, and, subsequently, the number of intercultural problems.

1985–1990

Another recession occurred shortly after 1985, and studies began of branching into new fields such as innovative applications of information systems, biotechnology, and new materials, as well as of the potential use of new environments, such as space, the oceans, and deep underground. The expansion of overseas production facilities proceeded apace, and the employment of foreign workers in Japan, legal or illegal, became a widely discussed social phenomenon.

1990–1995

The economic bubble burst in 1992, and economic growth of zero to a few percent became common. The Japanese overseas subsidiaries established during the previous periods were critically screened, leading to either withdrawal or reduction in scale of operation. Japan was the only low-growth economy among the booming Asia-Pacific economies. Sexual harassment became an issue at some of the Japanese subsidiaries in the United States.

1995–2000

A series of measures were undertaken to reform the industrial structure of Japan; however, it was not until 1999 that signs of substantial economic recovery could be seen in Japan. In 1997 and thereafter, financial and currency issues surfaced and became serious in Asian economies and Japan. The supply-demand gap in the Japanese economy was estimated to be over 30 trillion yen as of 1997, and a maximum of only 2 to 3 percent economic growth could be expected even with implementation of measures for expansion of domestic demand and deregulation.

Many Japanese companies became international, and some of them became globalized in various degrees through the stages of evolution described in Chapter 1 and Table 1.5. Many of these companies have adopted western-style corporate rules, systems, job descriptions, accounting, and annual reports.

SOCIETIES IN CONTRAST

An old Japanese proverb says, "Don't wipe another man's face with your duster." This means, do not say things to another person that he or she is disturbed to hear, or otherwise behave in a deliberately offensive manner. Courtesy in Japan dictates avoiding saying something to someone which would put him or her in an embarrassing position, or causing other difficulties, even in the competitive world of business. The polished Japanese strives for harmony *(wa)*.

Europeans and Americans, on the other hand, believe in claiming personal rights in debate, regardless of the potential for embarrassment or irritation of either side. This style lends itself to confrontation, something most undesirable in Japanese society. Similar western manners and ways of thinking have been propagated throughout the world for the last 400 years, while during the same period Japanese ways have been confined to Japan.

Generally, when Japanese work in close contact with foreign business people, they try to behave according to their counterparts' culture and manners, so as to give the maximum impression of compatibility. As a matter of principal, those who wish to do business in a foreign country must follow the laws, regulations, and culture of that country. Therefore, the Japanese expect this effort of those who come to live and work in Japan. Foreign-born engineers coming to work in Japanese companies and universities should expect to find cultural differences, learn what is expected in Japan, and try to follow the rules and manners of the Japanese as far as possible.

SKETCH OF A TYPICAL JAPANESE COMPANY

Conditions of Employment

Japanese workers have no specific employment contract with their employers, except those who are working after retirement or on a temporary basis. However, internationalized companies have begun to provide written contracts for foreign employees as of the early 1990s. Instead, at the time of their hiring, employees are requested to submit oaths with a letter from a guarantor. Companies give entrance examinations to applicants from universities or graduate schools six months prior to their graduation. Companies issue regulations which they expect everyone, including foreign employees, to follow. The company where you work may not have an English version of its regulation*, which will be troublesome but personnel department staff will explain them for you; you can always ask your colleagues if you need more concrete explanations.

Job Description

The organization chart when rendered in English will probably seem easy enough to understand at first glance, but sometimes the actual jobs employees and managers do will differ considerably from jobs in your country having the same names.

For instance, *supervisor,* in the United States, means a person who watches subordinates, guides them in performing their jobs as expected, evaluates their performance, bears responsibility for safety, and sees to it a large enough

---

* An increasing number of international Japanese companies have them.

workforce is on hand to meet deadlines. Supervising subcontractors is deemed a duty of a field engineer. However, in Japan, a supervisor has more responsibilities than this. He or she does all the above jobs plus those of field engineers, such as subcontract relations, quality control, and progress investigation.

Laborer in the United States suggests a worker who performs unskilled physical work. Such a person does material handling, cleaning, basic setup and teardown of equipment in the field, and digging, for example—but not derusting work with wire brushes, or painting for rust prevention, for example, which are done by apprentices. In Japan, however, the "unskilled" work a "laborer" is expected to do might include derusting, painting, or sawing logs for temporary fences.

Autonomous Activities

Engineers, plant operators, clerks, and even supervisors are expected to keep machinery, platforms, field sites, and offices clean on their own as they deem necessary, without instructions from their managers. It is the duty of any employee to keep the work area clean at all times and a fundamental principle of the philosophies of total productive maintenance (TPM), the zero defect movement (ZDM), and total quality control (TQC), all of which are carried out by small autonomous groups.

Autonomy is the watchword in Japan. Employees in Japanese companies make it a point of pride to do their jobs on the strength of their own motivation without instructions from managers. Managers often ask for opinions and suggestions from their subordinates when making certain decisions; they sound out the concerns of their subordinates on issues, and expect informed cooperation from them when the decision is finally taken. Managers do have their own opinions, but they like to follow this procedure in decision making.

Japanese are not as shy as they appear to westerners, but they respect modesty. In meetings, therefore, it is not polite to state one's opinion without permission before one's superiors have expressed theirs. This comes from the Confucian teaching "elders first, juniors second."

The Simplest Way to Fit into Japanese Culture

A common and effective form of communication in Japan is *nemawashi*, which means laying the groundwork for obtaining the agreement of one's companions, a kind of behind-the-scenes maneuver as stated elsewhere. It is a form of communication between parties outside the normal flow of information via management. It is not always appreciated in a western organization, as it bypasses the chain of command.

A majority of those who go to foreign countries to live and work suffer frustrations. They are working within a different culture, have difficulties with communications due to the language barrier, eat unfamiliar foods, and must accept a new style of living. These shocks can be eased greatly by having the proper attitude, however, which it is the intent of this book to foster.

The foreign employee is urged to feel free to consult colleagues on all questions about job procedures, evaluations, promotions, and raises if written versions of the rules are not available in English.

Finally, we would urge all who come to Japan to make a sincere effort to master the Japanese language, since this is the most effective way to feel at home with your fellow employees and with Japanese culture in general.

8

Features of Japanese Industries versus U.S. Industries as Reflected in Negotiations

Hiroshi Honda and Robert Latorre

INTRODUCTION

There are many similarities and differences between Japanese and U.S. industries. Similarities naturally appear in the industrial products, services, and basic technologies, and in recent years, they have been appearing in the overall production and service organizations. However, when the day-to-day operations of Japanese and U.S. industries are compared, we find a number of aspects which reflect the contrasts arising from differences in history and culture.

JAPANESE BUSINESS CULTURE

Japanese businesspeople tend to avoid using the classical western style of speaking out at first, until they are well acquainted with the other party and judge the time to be right. Such a style of discourse could turn off the Japanese, who value vague communication and group harmony over individual opinion. A decision is preferably made as a result of group discussions, and is seldom "won" in a Japanese organization unless it is truly necessary to do so. Japanese business has been based on long-term relationships and trust, as can be seen in Toyota's procurement of auto parts from its group *(keiretsu)* companies. However, this practice is changing, as can be seen in

Nissan's recent changes in procurement of auto parts from new companies outside of its keiretsu system. It is accepted in Japan that if two parties trust each other, detailed negotiations on schedules, payments, and other business formalities are dispensed with. So contracts, invoices, and product specifications may not be cited for conducting established business relationships in Japanese negotiations. Even if a written contract exists, the Japanese will not examine it in front of their business partners, to avoid implying any lack of trust, but they would certainly do so after the meeting. This style extends to the Japanese businessperson who builds relationships by giving presents to important customers.

JAPANESE INDUSTRY APPROACH

Japanese industry and Japanese business share a number of similarities in their way of negotiating. Japanese negotiations sometimes are not set on a timetable, but rather are characterized by persistence toward a long-term goal, investment of resources in preparing the way, and privately collecting information. Like the Japanese business style, it sometimes appears abstract and very personal. Japanese liken their negotiations as conflict resolution to the old style of an Edo period ballad called *naniwabushi* originated in Osaka. The negotiator lays the groundwork, reviewing the mutual benefits of the business and personal relationship in a prologue called *kikkake* and then turns dramatically to the crisis and an account of the reasons for the problems. In the "finale," the negotiator casts the group as the tragic victims needing concessions from the other side for survival. In contrast to the western idea of cause and effect, blame for the crisis is not always directed toward the other side, but toward fate. This difference in accountability could be seen in many past cases of negotiations between U.S. and Japanese industries. (Japanese industry is now getting used to U.S.-style negotiations, while some of the established Japanese companies already feel comfortable with the moderate European-style negotiations.)

The vague role of the Japanese boss in many negotiations and corporate developments is based on a Far Eastern strategy to succeed using intellect rather than power. This approach gives Japanese industry room to initiate new ventures and take advantage of unanticipated market development, as illustrated by the Walkman's appearance in the 1980s. Unlike western industries, which try to attack problems and make rapid decisions, Japanese industries tend to linger and wait to make decisions until things are more favorable. Japanese engineers focus on product quality and leave concern for the product market and profits to upper company management, which in turn direct the issue to marketing and other relevant departments for its feasibility. In Japanese industry, it is important to first develop customer service rather than sales. Consequently, corporate culture aimed at fulfilling customer needs is an important perspective in Japanese industry.

CONTRASTS IN U.S. AND JAPANESE INDUSTRIES AS HIGHLIGHTED BY A NEGOTIATION

Japanese may enter negotiations with westerners expecting a stressful time due to language and cultural misunderstandings, as well as their disdain for confrontation. On the other side, westerners have problems with being unable to deal with the Japanese, which creates awkwardness and discomfort on their side.

Japanese industrial teams work as a cohesive unit under one spokesperson. The Japanese spokesperson is not always the key decision maker. Usually, the decision maker is away until the negotiations are ready to conclude. Such teamwork is the Japanese' strength in the face of aggressive actions. Also, there will be no verbal or nonverbal feedback such as a nod or smile to acknowledge their counterparts' messages.

Throughout the negotiations, the Japanese side will reveal little of its strategy, or even who their decision maker is. The other part is that the Japanese side will want to deal with the broad agreement first and then settle the details. Japanese also handle every business problem on a face-to-face basis, not by telephone, fax, or mail, if they are serious about the business.

The American legalistic or lecturing style of behavior is sometimes offensive to Japanese. Verbosity is often seen by the Japanese as a personal problem and consequently as a barrier to effective communication. It is necessary to be ready to apologize for unintentional offenses, or rudeness, to restore the dialogue.

A meeting between a Japanese and western team may proceed as follows. The Japanese side will consist of four to six people. The most senior is a company director (often a member of the board of directors). The western side will be a team of two or three: a senior vice president, his or her assistant, and/or a technical manager. The negotiation meetings will go on in proportion to the extent of the negotiators' relationship. The longer and more cordial their relationship, the shorter the negotiation meeting.

Negotiation starts with a 20-minute opening or prologue from each side. The Japanese will stress their plans and how this venture will enhance both sides. The western side will develop the rationale for the anticipated results. One side will then mention price, which is followed by a counteroffer from the other side. For commodities and raw materials, there appears to be a lot of volatility and bargaining, especially when the group is from the Kansai-Chukyo (Osaka-Nagoya) area. It should be noted that people in that area in a sense enjoy the process of bargaining. For companies from the Tokyo-Yokohama area, there is less emphasis on the bargaining. In the discussion phase of the negotiation, areas such as customer service and technical support are focused on.

Two situations typically arise in these negotiations:

1. The Japanese industry will attempt to modify the contract terms to reflect changes in its circumstances, while the U.S. industry will attempt to forecast and include the worst-case situation and treat things more legalistically.
2. The U.S. industry will often abandon a losing activity, while the Japanese industry will have a tendency to persist even if there is no real hope of success, if it can see possible by-products and/or possible derivative contributions to the entire business.

Finally, after several critical time-outs and consultations, the two sides hopefully reach an agreement which is satisfactory to both sides and builds a stronger business partnership.

CONCLUDING REMARKS

Unique U.S. and Japanese styles of industrial communication and negotiation provide us with the following four U.S.—Japanese contrasts one should be aware of:

1. Japanese teamwork approach versus U.S. individual approach
2. Japanese vagueness at the initial stage versus U.S. direct style of communication
3. Japanese need to collect information without drawing conclusions versus U.S. use of limited information to develop the business/industry plan
4. The U.S. focus on short-term business goals as compared with the Japanese long-term approach.

Keeping these four contrasts in mind will enable the reader to pass beyond the stereotypes and misunderstandings that often occur in U.S.—Japanese business relationships.

9

University Education in Japan: A Personal View

Robert M. Deiters

Many readers of this book will be engineers just arriving in Japan (or only contemplating coming to Japan) to work in industry. But even for those readers, who have no intention of teaching in a Japanese university, my comments in this chapter should illustrate some important points. It is only discouraging and frustrating to wail about the unfamiliarity of the customs or your inability to work in ways with which you may be uncomfortable initially.

Whether you are in an industrial or an academic situation, look first to see what cultural and human factors in your new milieu promote cooperation, a sense of belonging and participating, creativity, and growth. Use these. They are the real strength in the Japanese way of doing things.

EDUCATION IN JAPAN

In 1980's, higher education in Japan has drawn criticism both within Japan and from foreign observers: the curriculum is not tightly organized, observers say; little work is required of the students, and anyone can expect to graduate once he[1] gets in; and many students spend their time at side jobs, traveling, or

---

[1] Throughout this chapter I will commonly use *he* and *his,* because, of the hundreds of students I have taught over the years, most were young men, with only a sprinkling of women among them. Consequently, I ask the reader to allow me to use these masculine pronouns, because I can express my actual experience more naturally in this way.

at work on either their own personal projects or those of the numerous student clubs. Some critics go so far as to describe the Japanese university as a large—and for the parents, expensive—"recreation center."

It is now some decades since I introduced myself in halting Japanese to my first freshman class. To some extent I have to agree with these criticisms, and—what is even more embarrassing—I have to admit that I have been part of the problem in over 40 years of teaching in Japan. But as is often the case when the works of Humankind are judged, the critics have been more adept at pointing out the faults of the Japanese university than at pointing out its real strengths and potential.

In 1954 I was 29 and the members of the A-4 class (each class of 40 to 50 students becomes a social unit with a unique identification number) were 19-year-old freshman—that was my first class at Sophia University (in Japanese, Joochi Daigaku). Once in a while the "A-4 Old Friends," now in their mid-sixties, gathers. I and another former teacher were invited on their annual overnight trip to a hot spring resort. Although I had taught the class for only two hours a week in the freshman year and then had left teaching for some years to complete my own studies, they have kept friendly contact with me over the years. As a foreign teacher in Japan I sometimes experience frustrations, but the deep satisfaction of my friendship with students and graduates far out-weighs the frustration. I have some critical opinions about university education in Japan as I have experienced it, but deep down I know that a system of education that can form a group like the A-4 Old Friends is basically sound.

Beginning about the late 1970's, when productivity of Japanese manufac-turers began to draw the notice of Americans, articles and books about "Japanese-style management" came out one after another. By now, though, the more perceptive of Americans have already realized that the labor-man-agement relations in Japan are rooted in Japanese cultural and social pat-terns that cannot be simply translated to America. Improvement by imitation will not succeed.

Ever since the postwar Allied occupation of Japan changed the educational system to match—outwardly—the American model, some visiting academics, particularly Americans, have criticized the Japanese universities because they do not do what the better universities in the United States are doing. Such comparisons are useful, I believe, but imitation, once again, is an impossible and unrealistic goal. Just as a corporation in the United States cannot be man-aged in a purely Japanese style, neither can a Japanese university educate its students with the curriculum and classroom style of MIT, for example. In Japan the university, too, is already a well-established social institution, rooted in a cultural milieu which has developed continuously for centuries. An institution, like a human being, can improve only by making the most of its own special characteristics. Demanding from an institution a new style of behavior that has no roots in the culture and is alien to the people who work in it will lead to frustration. Therefore, in my own critique of university education in Japan I have first restrained my tendency to compare it with the system in American universities I have known. Then I have tried to observe where and how the Japanese university—at least the one where I teach—is successfully educating students, and I will try to offer a few suggestions for extending its success.

With the passage of years—it was 1952 when I first came to Japan—I have become convinced that the two experiences which best succeed in educating the typical university student are participation in a student club and the graduation research project. After expanding a little on why I judge these two to be so successful, I want to offer a few suggestions on how university education might be restructured somewhat to build on these strengths.

Earlier in my teaching career I found myself telling students, especially the incoming freshman, that it was dangerous to get too involved in a student club or circle. "You won't be able to spend much time or energy on studies if you get caught up in one of those student clubs or circles!" Fortunately few students headed my advice. Students have a healthy, almost instinctive sense that it is good for them to belong to a student club or circle. And when I think back on the students I've known, and especially after I see how they have developed in society after graduation, I must recognize that the shaping they received through the club was an important formative influence—perhaps one of the most important.

On the Sophia University campus, student clubs and circles proliferate like mushrooms in a rain forest. There are over 300 clubs and circles: for photography, ballroom dancing, rugby, off-road bicycle touring, mandolin ensemble, volunteer service to the handicapped, Andean folk music, movie appreciation, English-language drama, and English-language debating, and more than 30 circles for tennis—a rich profusion of youthful energy and developing talent. Largely outside the direction of the professors, these young adults, already at the age when personal aptitudes and intellectual acumen are reaching full flower, are molded in a social atmosphere which encourages initiative, allows scope for creativity, and provides just the right amount of criticism and coaching. I am constantly amazed at the high goals the clubs set for themselves. The camaraderie of the club or circle encourages and rewards committed effort while at the same time helping beginners progress beyond their first imperfect attempts.

Years ago I was amazed to learn that, for example, the main tennis club, which, like the "varsity" in an American university, represents the university in intercollegiate matches, admits as full member first-year students who have never yet held a racquet. If only the student commits himself to practice seriously and to take full responsibility as a member of the club, he is provided with coaching and encouragement to progress. Students who could play no musical instrument when they entered sometimes perform in a first-class student orchestra before they graduate. An incoming freshman who can hardly speak a simple sentence correctly in English is able to debate in English a few years later. There is in these clubs a secret of successful education which we teachers in universities should learn. I have corrected my judgment of what I, at first, considered to be just "playing around."

In my view of education, a young adult can only educate himself; his own self-initiated activity is all-important. The role of the teacher is to give some guidance and motivation for such activity: to coach and encourage, to correct, and to challenge the student to extend his talents to the fullest. The young adult student should be given as much autonomy as possible to develop his personal talent, but he also must be called to stick to his choices, to commit himself to what he has undertaken. Also, for sound development, whether in tennis or in mathematics, he needs friendly and frequent coaching. But he should be helped to grow the point where he has the power to evaluate his own efforts and impose the needed discipline on himself.

The club gives the student what professor Chie Nakane, a cultural anthropologist, has called the *ba,* the place and "frame," for his life and activity. Nakane has developed in her two books *Human Relations in the Vertical Society,* and *The Dynamics of the Vertical Society*[2] the hypothesis that the primary

2. "Human relations in the vertical society" is my translation of the title of Chie Nakane's book *Tate-shakai no Ningen Kankei,* Kodonsha, 1967. Although not a translation, Nakane's book, *Japanese Society,* University of California Press, Berkeley, 1970 (also Penguin, New York, 1973), seems at a cursory examination to cover the same matter. The Dynamics of the Vertical Society is my translation of the title of Nakane's book *Tate-shakai no Rikigaku,* Kodansha, Tokyo, 1968. As far as I know there is no English translation of this book.

social unit, or "molecule," for all Japanese is the small social group within which vertical (senior-junior) human relations are far more important than peer relations. Japanese, according to her, give their allegiance as well as their time and energy primarily to only one group; one cannot be equally a member of several groups. In the club the student finds his frame, where he has his place in this small "vertical society." Here he finds familiar vertical human reactions among friends both older *(sempai)* and younger *(koohai)*. Here he feels at home and can engage in creative activity in an environment of common interest and shared responsibility.

The other very successful educational activity on universities is the graduation research of fourth-year students. In the electrical engineering department where I teach—and the practice is quite similar in other universities—students, just before the beginning of their fourth year, are shown the list of research themes or projects that each individual professor proposes for senior student research. Trying to satisfy the desires of the students as much as possible, we assign about six senior students to each teacher. During their entire senior year the students assigned to me are "members" of my laboratory. A young faculty assistant also has his desk in my laboratory. There are also some graduate students who have been with me for two years or more. This group naturally forms the type of small group which Professor Nakane says is so natural and congenial to Japanese. This group has an ideal ba, the laboratory, the scene and the center of the group's activity and social life. The group is bound together by friendship, by healthy older-younger "vertical" relationships and by a senior leader, the teacher, who, in Nakane's model, symbolizes and maintains the unity and smooth working of the group.

In the atmosphere of the laboratory, I have observed, the students immediately feel at home: relaxed, accepted as individuals, incorporated in to a primary "vertical society." This kind of ba ensures the essential human conditions for personal initiative and creative activity. In Japan it is only the very rare "loner" who can even get interested in studies which are not integrated in to such a small-group atmosphere of friendship and cooperative effort. The goal, the theme of an individual student's graduate research, as in any true research, is usually not sharply defined, and so the student is challenged to creative self-determination. Also, commitment to his autonomously selected project is demanded of him. The group, by its friendly, relaxed atmosphere, encourages him and sustains him through the long, difficult efforts at research. He is not alone. And yet he has personal responsibility for selecting his goals and the means by which he will attain them. Finally, at every step of the way, he is corrected (mostly by imitating his seniors and emulating his peers), frequently coached, and his work evaluated.

Although the professor oversees the whole process, much of the education goes on without any need for him to be directly involved. The professor is free to plan the overall direction of the research and to advise at crucial junctures. In this atmosphere students who, in their class work, seemed mediocre and colorless blossom into energetic and creative researchers. I have had students who significantly extended the research I had done in graduate school. One student, who among us teachers had the reputation of being a poor achiever (it was already his sixth year in the university), found and corrected a mistake in a long calculation I had thought to be correct. I had not asked him to check the calculation: he had done it on his own initiative just to understand better what he was doing in his own project. Some of the final undergraduate research reports I have seen are of a quality fit for publication as research papers.

Almost every year I am amazed at the rapid progress and great ability shown in this graduation research. And yet, I am almost equally amazed at the passiveness and the lack of interest in the lecture courses and poor performance on examinations we professors give at the end of each semester. Are these the same students? What does the success of this graduation research tell us about how we might improve the rest of the process?

One of the greatest defects in the Japanese university is, I believe, that the students are "talked at" too much by professors. They have too little time and too little guidance and encouragement for their own personal study and practice. Too many hours siting passively in the classroom starts a vicious cycle, for the student has too little time and energy left for personal study. A student in engineering spends an average of six hours a day in the classroom or laboratory Monday through Friday, and possibly two to four more hours on Saturday. Also, students at Sophia University in Central Tokyo commute an average of two hours a day on crowded trains and busses. Unless a student trains himself to "four-hour sleep" (the title of a book written by one of my colleagues), he cannot possibly get in three to four hours of solid private study each day. Many engineering students take almost no time for private study on an ordinary class day.

The result is that the student cannot personally assimilate what he hears in class. At the most, he carefully notes down what the professor says in order to review it the night or so before the examination. The teachers, on the other hand, realizing that the students have little time for personal study, tend to explain every detail in the classroom. The pace drops. Also the professor finds it almost impossible to promote individual activity by assigning projects and reports, because he has too little time and help to go over the papers and give feedback to the students. In Japan, where younger people learn so much from their immediate seniors in the small group, what a waste it is not to have the older students correcting the reports and homework of the younger students! But the greatest problem with the system of lecture courses is that the student belongs to no group, has no ba centered on curriculum-related study.

Thanks to the entrance examinations system (called "examination hell" by the young), the typical engineering student enters the university with very finely honed skills in mathematics and exact knowledge of what might be called "factual" physics and chemistry. However, he also brings along a habit of studying only to master facts and acquire computational and problem-solving skills. From the very beginning the university should break that pattern to guide the student toward a more self-directed and critical method of study. In order to do this the strengths of the small social group and the vertical society should be brought into play.

Therefore, I propose that, immediately after entering the university, each student be incorporated into a small group (call it a "seminar-club") associated with one of the professors and the older graduate students under his direction. In my own department, for example, such a group consists of about 24 students, six from each of the four undergraduate years, together with about six graduate students. The group would hold some seminar-style meetings, and each student would have his turn to make a presentation and have his work reviewed. The seminar presentation of undergraduate students would, for the most part, concern projects (homework, if you will) set by the professors giving the general lectures. In this way, the work of the students would be examined and corrected in the seminar group before being given in for final form to the professor in charge of a particular course. (I can hear an American reading this say, How will you evaluate the personal effort and comprehension of individual students if they get help from others? My answer is that grades are just not that important in Japan. There are many other ways of evaluating each student—for example, by his contribution to the group.) The young student could also begin in some way to participate in the research of the group, especially during the long spring and summer vacations, which are otherwise lost—for academic purposes—during the first three years of the university.

This "seminar-club" would make it possible for all of the students to help in one another's education. The professor, the leader of the seminar-club, with his experience and larger view of the field, could contribute his best by not being overburdened with the many details that go along with direct supervision of each student's independent activity.

Then too the number of hours of lectures could be reduced. The lecturers could go ahead at a faster pace, satisfied that the students could master the material in private study and with the help of their seminar-club. The teachers would no longer have to presume (as they do now) that when the student is not in the classroom or laboratory, he is very likely with his club planning and preparing to cross the Sahara Desert on off-road bicycles next summer.

10

Teaching in English at a Japanese University

Robert Latorre

JAPAN'S EDUCATION TRADITION

Education in Japan has roots in its Confucian and Buddhist heritage, which stresses hard work, diligence, and perseverance as the means to success in education and other aspects of life. This is balanced by a recognition of the important responsibility borne by teachers, parents, and schools to awaken a desire to try. Motivation in learning and character is perceived in Japan as being shaped by teachers and influenced by the school environment.

Dr. E. Herrigel, who taught in Japan before the Second World War, wrote:

> The Japanese pupil brings with him three things: good education, passionate love for his chosen art and uncritical veneration of his teacher. The teacher-pupil relationship has belonged since ancient times to the basic commitments of life. ... The teacher considers it his first task to make [the student] a skilled artisan with sovereign control of his craft. The pupil follows out this intention with untiring industry. ... How far the pupil will go is not the concern of the teacher and master. Hardly has he shown him the right way when he must let him go on alone. The master exhorts the pupil to go farther than he himself has done and to "climb on the shoulders of his teachers."
>
> It is this educational tradition which sustained the Japanese students who went overseas and created conditions which are the basis for Japan's University education.[1]

---

[1] E. Herrigel, *Zen in the Art of Archery,* Vintage Books, New York, 1971, pp. 41–55.

EARLY ROLE OF FOREIGN ENGINEERING INSTRUCTORS _____

The arrival of Perry's fleet set into motion the events which led to the end of Japan's isolation and to its modernization. In order for the Meiji leadership to realize the slogan "Fukoku kyouhei" (a wealthy and strongly armed nation), it was necessary to undertake a major import of technology. This included hardware, technical methodology, and the employment of more than 800 foreign technicians and teachers between 1879 and 1890. In 1873, Professor Henry Dyer and his British engineering instructors established the School of Engineering in Tokyo, which later became the Imperial College of Engineering (1877–86) before being merged with Tokyo Imperial University in 1886.[2] The initial success of these foreign technologists in Japan is part of the background for the change in Japanese law in 1982 that stipulated that foreigners may be employed in national and municipal universities under the same conditions of employment as Japanese.

EMPLOYMENT OF A FOREIGN PROFESSOR _____

Formalities

The 1982 Japanese law provided that a foreigner may obtain the position of professor, associate professor, or lecturer and may participate in faculty meetings with all privileges such as voting, just as Japanese may. The term of employment is two to three years, which may be extended by application. In the 1986–87 academic year, I was hired by the University of Tokyo Mechanical Engineering Department as an associate professor. This was the first foreign engineering professor appointment since the appointment of Professor Charles West (1847–1908) appointment in the Meiji era.

The first stage was Professor Hideo Ohashi's creating a position in the mechanical engineering faculty. Initially the idea of a three-year

[2] R. Latorre and K. Hongo, "Relevant Features from the Benchmark Period of Japanese Engineering Education (1867–1900)," *International Journal of Applied Engineering Education*, Vol. 5, No. 3., 1989, pp. 341–50.

appointment was difficult, because of the inability of my U.S. university to grant more than a one-year leave of absence. Subsequent appointments to this position have been awarded to U.S. professors seeking employment.

The second stage involved completion of the documentation described in Table 10.1. The necessary documents were translated into Japanese and presented for review and approval (June 26, 1986) in a meeting of the mechanical engineering department. After the faculty of engineering registered its approval, the University of Tokyo applied to the Monbusho to complete the employment formalities and the 4-1-7 visa certification from the Ministry of Justice. This was completed after my arrival in Tokyo when I signed the *senseisho* to work for the Japanese government on September 1, 1986.

Table 10.1   Documentation Required for Employment of Dr. Robert Latorre as Associate Professor, Mechanical Engineering, University of Tokyo, August 1986–September 1987

Item	Documentation

1. Curriculum vitae including personal details; educational history from primary school to graduate with starting and ending dates; professional experience including work history; academic degrees with dates and issuing institutions.
2. Certificates of primary school, junior high school, high school; certificates of B.S.E. and M.S. from University of Michigan. Note: My M.S.E. and Dr. Eng. from University of Tokyo were already on file at the University of Tokyo.
3. Certificates of professional appointment as an Associate Professor (University of Michigan) and Associate Professor (University of New Orleans), and Certificat de Travail–Stagiaire, ARMINES, Paris, France.
4. Health certificate including results from a chest X ray.
5. Two photographs.
6. Signed declaration that I am neither interdicted nor incompetent, and have never been sentenced to imprisonment.
7. Letter from UNO College of Engineering dean certifying that I am entitled to a one-year leave (September 1986–September 1987) for the purpose of teaching at the University of Tokyo. In addition, the university regulations governing such leave of absence.

Activities

In June 1986, Professor Ohashi wrote, "Your duty at our Department will be mainly teaching in English. I expect you to teach part of the advanced course of fluid dynamics and the graduate seminar for fluid dynamics. You will be asked also to take part in the Exercises in Mechanical Engineering 1 and 2 (Junior Year, Winter and Summer Semesters). All three teaching duties will require 2 or 3 frames (1.5 hours per frame per week)." This broadened into all the activities shown in Table 10.2. These activities included student supervision and participation in laboratory activities, as well as mechanical engineering, faculty of engineering, and graduate school meetings and committee assignments.

Results of the Engineering Lectures in English

To measure how they were receiving the English engineering lectures, the students filled out blank survey forms, which they did not sign. The students rated their ability on a scale of 1: poor (0–40 percent), 2: average (40–60 percent), 3: good (60–80 percent), and 4: excellent (80–100 percent) on the following three points:

1. Follow the lecture
2. Hear the spoken English
3. See/write the English notes on the blackboard

My impressions and discussions with my Japanese colleagues indicated that a Japanese student's English ability is uneven. A typical Japanese student had a reasonable ability to read and write English (item 3) but only a limited facility in hearing and speaking English (item 2). By surveying the students just at the end of each lecture, I hoped to clarify to what extent the good visual and poor aural English ability impacted comprehension of the technical English lecture (item 1).

While the majority of the students had more than seven years of English training, they indicated problems with following the engineering lectures in English. This is shown in Tables 10.3a and 10.3b. At that time, there was no teacher evaluation system in Japan. This survey represents a useful sample of the difficulty these Japanese students had with English lectures. I found it was difficult to motivate any student participation.

Table 10.2 Activities of Associate Professor Robert Latorre, Mechanical Engineering Department, University of Tokyo, August 1986–September 1987

Activity	Code	Frequency	Purpose
Lab meeting	A	1 × wk × 3 hr	Schedule research, monitor B.S.E, M.S.E., and Dr. Eng. thesis
Thesis	A	1 × wk × 2 hr	Thesis supervision
Research	A	3 × wk × 2 hr	Cavitation research on noise and inception
Dept. meeting	B	1 × wk × 2 hr	Conduct department business
Associate prof.	B	1 × wk	Get together lunch (develop English correspondence PC disk*)
Lecture	B	2/3 wk × 1.5 hr	ME lectures (Fig. 2/Table 3)
	B	2 × wk × 2 hr	Course development/lecture preparation
ME exer.	B	1 × wk × 3 hr	I—gear box design
			II—cavitation studies/English usage
Field trip	B	1 × wk × 3 hr	Visit pump factory with 30 students
Committee	B	1 × month × 3 hr	IBM–Japan–Univ. of Tokyo ME Dept. Partnership for Computer Usage
Committee	B	1 × month × 1 hr	ME Library Committee (introduction of video,* Prof. West video*)
Workshops	B	2 month × 3 hr	I—technical English presentation*
			II—use of OHP in video presentation*
Exams	B	4wk × 4 hr	a) Univ. of Tokyo entrance I and II
			b) M.S./Dr. Eng. entrance
			c) Thesis exam
ME brochure	B		Develop ME brochure with Dr. Inaba*
Fac. meeting	C	1 × month × 3 hr	Conduct engineering faculty business
Committee	C	1 × wk × 2 hr	Revise engineering catalog
Grad. fac.	D	1 × month × 3 hr	Policy and examinations
Endowed chair	E	3 × 2 hr	Obtain materials on U.S. endowed chair
Speech	E	1 × 2 hr	Topic: use of video for efficient time use
Panel	F	—	Topic: U.S. engineering education
Panel	F	—	Topic: future of Japanese science and industries
Workshops	F	1 × wk × 4 hr	Technical English presentation*
Paper	G	—	U.S. ocean engineering research (chair session)
Committee	F	1 × wk × 3 hr	Translation of ASME Code into Japanese

Key: A, Prof. Ohashi–Prof. Matsumoto–Latorre laboratory; B, Dept. of Mechanical Engineering; C, Faculty of Engineering; D, Graduate Faculty of Engineering; E, University of Tokyo; F, Japan Society of Mechanical Engineers, G, University of Tokyo, Institute of Industrial Science.

* Initiated by Dr. Latorre.

Table 10.3a   Student Feedback on English Lectures, 200071 Fluid Mechanics I, Second Year Students, Monday 10:00–12:00
Replies of 61 students (12/1/86), 46 currently studying English.
Number of years of English training: 6–8 years: 48 students
Over 8 years: 14 students

	Number of	Follow the Lecture				Hearing				Seeing/Writing			
Date	Students	1	2	3	4	1	2	3	4	1	2	3	4
11/10	67	23	31	11	2	26	35	4	2	18	31	17	1
11/17	48	16	17	13	2	17	20	10	1	6	18	21	3
12/7	62	21	29	12	0	18	34	10	0	8	28	26	0

Key: 1, poor (0–40%); 2, avg. (40–60%); 3, good (60–80%); 4, excellent (80–100%).

In the 200071 Fluid Mechanics I courses, I assigned homework. In the class of December 1, 1986, I asked for a volunteer to write the solution on the blackboard. No student raised a hand. So after a pause, I wrote out the solution. Several students then wrote notes to me on the survey form to indicate why no one would volunteer. Here are excerpts from one reply:

> In the present curriculum of Japan, virtually 100% of university students have studied English from junior high school. So we have all had at least $7\frac{1}{2}$ years of study.

Table 10.3b   Student Feedback on English Lectures, 774–22 Advanced Fluids Engineering, Graduate Students, Wednesday 10:10–12:00
Replies of 25 students (10/29/86), 2 in English class. Number of years of English training: 6–8 years: 13 students
Over 8 years: 8 students

	Number of	Follow the Lecture				Hearing				Seeing/Writing			
Date	Students	1	2	3	4	1	2	3	4	1	2	3	4
10/15	20	0	6	12	2	3	10	6	1	1	4	10	5
10/22	23	0	7	13	3	5	12	4	2	0	9	10	4
10/29	25	2	14	9	0	5	10	8	2	1	7	12	5

Key: 1, poor (0–40%); 2, avg. (40–60%); 3, good (60–80%); 4, excellent (80–100%).

Regarding the lack of response on the homework solution, the student continues,

> Please don't get disillusioned that no one has done it, sir. I expect that over half
> of us have solved it. It's just that we Japanese are sometimes too "modest"
> (speaking cynically!). It's strange, but it's something typical of Japanese (even in
> our university). Thank you for a nice lecture.

DIFFERENCES BETWEEN U.S. AND JAPANESE UNIVERSITIES ___

Table 10.4 draws a number of comparisons between Japanese and U.S. universities. The Japanese universities represented are the larger national and private
universities where each engineering professor chairs a laboratory. The assistant
professor joins an existing laboratory and assists the full professor in conducting research, supervising students, giving lectures (usually in team teaching),

Table 10.4   Comparison of Japanese and U.S. Universities

Factor	Japanese	U.S.
1. Academic year	2 terms Summer: April to September Winter: October to March	$2^1/_2$ terms Fall: September to December Spring: January to May Summer: May to August ($^1/_2$ term)
2. Entrance	Once a year in April by exams in January–February	Usually in September, other time possible
3. Graduation	Once a year in March Completion of graduation thesis	December, May, August at completion of credit
4. Each course	1.5 contact hours per week, by 2 or 3 professors	3–4 contact hours per week, by 1 professor
5. Research components	More computer and experimental work	More analytical computations due to limited lab access
6. Laboratory	One affiliated with each full professor (similar to German chair system)	Typically 1–2 per department
7. Faculty salary	12 month	9 month
8. Research perspective	Long-range following senior professor's research	Long- and short-term following funding priorities

and serving on many committees. Promotion to full professor implies becoming the director of the laboratory.

The differences between Japanese and U.S. universities require attention to:

1. Schedule differences due to the poor match-up between the Japanese and the U.S. academic year
2. The need to interact with numerous staff and students in realizing research goals
3. A visitor's need to align his or her research with current laboratory research topics

Item 1 remains a problem, because of basic differences between the Japanese and the U.S. academic year. However, items 2 and 3 can still be resolved for maximum mutual benefit to host and visitor.

MAXIMIZING MUTUAL BENEFIT

The visit of an English-speaking professor is viewed with a wide range of expectations on the part of the Japanese hosts, students, and staff. To a large extent, the professor will be remembered for how well he or she was able to communicate with the laboratory members and impact their English-language ability. The hosts will strongly hope that students will show some tangible improvement in their ability to conduct lab activities in English. A two-stage approach has proved very useful in maximizing communication:

Stage I. Weekly group reading of a technical paper

1. Groups of four to six are given a relevant technical paper, and meetings are scheduled each week for one to two hours.
2. Each member will read the paper out loud for 10 to 15 minutes. Members will be assisted by the visitor in pronunciation and in reading mathematical formulas and technical papers. The group will have gained a smoother presentation style and a more comprehensive technical vocabulary.

Stage II. Biweekly preparation of a research project summary in English

1. Each member in the laboratory group will be asked to write a draft of a four-paragraph English summary of his or her research project.

2. After correction, the summaries will be edited into a report for use by the professor to introduce the laboratory activities.

3. Upon completion of the English summary, each member will expand his or her summary into a one- to two-page abstract to send as a proposal to a U.S. or international conference.

The benefits of steps 1 and 3 of stage II are that the lab will have an English summary of its activities and the visiting professor will have a clear understanding of the group's projects and future activities.

Follow-Up

Depending on the results from stages I and II, there are a number of ways to follow up to further enhance English-language communication:

1. Workshop on technical English presentation
2. Use of movie or television videos to improve English listening ability
3. Developing a training diskette containing examples of routine correspondence dealing with research, publishing a paper, arranging a visit, etc.
4. Preparing a joint state-of-the-art article on Japanese research
5. Preparing a proposal for future cooperation and exchange

A lot can be accomplished in the one- to two-hour weekly meetings. They will be a welcome break from the research work. In many cases, the visiting professor can propose such meetings at the initial stage of contact and will find it convenient and appropriate to pass out a schedule during the lab session.

CLOSURE: DEVELOPING BALANCE IN EXPECTATIONS _____

Over the six years I was a graduate student at the University of Tokyo, I met only a few foreign visiting academics, and these were people who were on sabbatical and had arranged to spend their leave in Japan. What impressed me most was how wide a spectrum of purposes and expectations they had. For some visitors the trip to Japan was a tour where, after a short period in the laboratory, they began to travel. (This was a period when the exchange rate was more favorable.) Some other visitors, on the other hand, disappeared into their laboratories and ignored everything but their academic work. Because these early visitors were few and far between, they individually left a profound impression on Japanese hosts.

I recall the frustration expressed by one French researcher who had planned a very extensive research project but found himself in a laboratory where the previous foreign visitor had spent most of the time away from the lab. The hosts had assumed the Frenchman had similar intentions, so the lab equipment was all being used by the students when he arrived in September, and he could only begin his work in December when they finished their thesis work.

There is no ready solution to avoid these misunderstandings. It is up to the sponsors of visiting professors to decide whether it is important to make the effort to match up the expectations of host and visiting professor, if not doing so could make either one unhappy or unproductive. Typically a visiting professor's and a host's feelings will revolve around expectations in the following areas:

1. Technical
   a. Extending individual research
   b. Exchanging information
   c. Developing new ideas for future work
   d. Visiting Japanese scientific and industrial complexes

2. Language
   a. Learning Japanese (visitor)
   b. Improving English (host)

3. Society
   a. Discovering Japan (visitor)
   b. Creating overseas contacts (hosts)

4. Travel
   a. Touring Japan
   b. Visiting the United States

Initially, the visiting professor's and sponsor's expectations regarding the short-and long-term results from the visit may not be clear. It is necessary to discuss your expectations with your host. You can then be more assured of a successful time as a visiting professor working in a Japanese university.

References

E. Herrigel, *Zen in the Art of Archery*, Vintage Books, New York, 1971, pp. 41–55.

R. Latorre and K. Hongo, "Relevant Features from the Benchmark Period of Japanese Engineering Education (1867–1900)," *International Journal of Applied Engineering Education*, Vol. 5, No. 3., 1989, pp. 341–50.

R. Latorre, "Todai Experiment: Internationalism, a Foreign Engineering Professor and English Instruction Results," *International Journal of Applied Engineering Education*, Vol. 6, No. 6, August 1990, pp.607–16.

From the Editor

The origin of Japanese higher education can be traced back to 828, when the Buddhist priest Kuukai established Shugei Shuchiin (literally, the Institute for Integrated Arts and Various Wisdom) in 828 in Kyoto to teach Confucianism to priests and, to ordinary people, both Buddhism and Confucianism. Kyotoites say this would be the first model for the Japanese university. Even though the institute did not last long after Kuukai's death, various other schools appeared and disappeared thereafter from time to time.

Professor Latorre was a pioneer in teaching classes in English at a national university in Japan, whereas private universities, such as the International Christian University and Sophia University, had already been holding classes in English for a long time. The University of Tokyo has changed its curriculum and other systems since the time of Professor Latorre's visit to conform to the model of a graduate school-oriented university, as have the other major national universities, and in keeping with a program of internationalization. Today, even Japanese professors teach graduate courses in English for both Japanese and international students in Japan.

Concerning faculty hiring of foreign nationals, Japanese national universities have recently begun to offer a modest number of tenured positions to foreign nationals. The amendment of the rule concerning hiring of foreign nationals, as reflected in the accompanying excerpt from the internal rules of the Kyoto Institute of Technology, has facilitated this trend. See Chapter 22 for details.

Hiroshi Honda

Rule for the Term of Foreign Faculty Hired
by Kyoto Institute of Technology

Enacted March 18, 1985
Amended September 30, 1988

(Object)

Article 1: This rule is to decide the term for professors, associate professors, and lecturers who are foreign nationals, hereinafter referred to as the foreign faculty, to be hired by Kyoto Institute of Technology, based on the Special Measures Law on Hiring of Foreign Faculty (Japanese Law Number 89 enacted in 1982).

(Term)

Article 2: The term for employment of the foreign faculty shall be three years, and reappointment shall not be prevented.

(Exception)

Article 3: Besides the aforementioned articles, the president of the institute can fix the term of employment for each individual in the case of special situations, based on discussion at the institute council.

Additional Clause: This rule shall be enforced beginning March 18, 1985.
Additional Clause: This rule shall be enforced beginning October 1, 1988.

December 17, 1998

Part of the rule concerning Term of the Foreign Faculty Hired by Kyoto Institute of Technology shall be amended as follows:

In Article 3, "the president of the institute can fix the term of employment for each individual" shall be amended as "the president of the institute can either fix or unfix the term of employment for each individual"

Additional Clause: This rule shall be enforced beginning December 17, 1998.

Kohsuke Kimura
President
Kyoto Institute of Technology
Unofficial translation by Hiroshi Honda

11

Family Considerations

Deborah A. Coleman Hann and Stephen A. Hann

When reviewing the original draft of this chapter, we found that it was essentially a 2200-word condemnation of high prices in Japan in the late 1980's and in the early 1990's. However, from the late 1990's to year 2000, the prices have become reasonably moderate. We have decided to reduce our original tirade to a single section and devote the rest of the chapter to describing what our life has been like in Japan. We will restrict this discussion to what we have personally seen while living at our own expense in the Tokyo area for almost two years. Families with generous large-company sponsorship and the attendant benefits would have an entirely different story.

JAPANESE PRICES

There is no point in repeating the familiar horror stories—although they were in large part true—about the astronomically high price of almost everything in Tokyo in the late 1980's through the early 1990's. A fairly reliable estimate for the cost of living in the Tokyo area in 2000 can be arrived at as follows:

1. Domestic goods and services cost about 1.5 to two times what they would in the United States.
2. Imported goods cost higher than what they would in the United States.
3. The fees and deposits required to lease an apartment total a minimum of five months' rent.

Using these estimates when planning for living within commuting distance of Tokyo will probably provide a reasonably accurate budget for a foreign engineer and an accompanying family. Of course, many goods, especially mass-produced consumer goods, and services are reasonably or lower priced in Japan, but the ones that are not tend to make the average costs adhere to these estimates, which are based on a few assumptions:

1. Minimal purchases are made at trendy boutiques and top name stores.
2. Most entertainment at top clubs and restaurants is done at company expense.
3. One's apartment is at least a half-hour train ride from the city's business district.
4. The international schools are excellent but are affordable only by company-sponsored families.

There is simply no upper limit on how much money you can spend in Tokyo; salaried Japanese and self-financed foreign workers have no choice and abide by these restrictions. While the prices still cause us some discomfort, by applying discipline to our spending, we can afford to live in the greater Tokyo metropolitan area.

Having brought this issue out into the open, we will not dwell on prices in the rest of this article, but they are always lurking in the background as a trap for the unwary.

HOUSING

Upon arriving in Tokyo in March, 1999, we originally rented an 84-square-meter apartment in Tamagawa Gakuen, about 30 kilometers west of central Tokyo (see Appendix 7). It had three bedrooms, a living room, a dining room, kitchen, and one bathroom. It was on the first floor of a two-year-old six story typical modern Japanese apartment building. It had carpeted bedroom floors, but no automatic dishwasher, and since we were first tenants, we had to buy and install heater-air conditioning units (in Japan, central heating and air conditioning is not commonly used). The dining room was in traditional Japanese style with *tatami* (woven straw mats) as floor covers and *shoji* (sliding paper doors) to separate it from the living room. It was a twelve-minute walk to the train station and a 45-minute express train ride to the nearest of the really important thoroughfares of Tokyo, Shinjuku. We found the neighborhood extremely pleasant and very much enjoyed the apartment, even though we picked it mainly because of the rent of 160,000 yen per month. The rents as of 1999–2000 in Tokyo Metropolitan Area and its vicinity are shown in Appendix 6. These rates are much lower than those in the early 1990's. The problem was that our complex was built into a hillside and the walk to the station meant climbing exactly 100 steps and then descending a steeply graded street. In temperate weather, which exists most of the time here, this was a minor inconvenience. But during the two or three months of hot, humid weather, we would arrive at the station soaking wet from the exertion of the walk. Also, it was a 1 $1/2$ hour trip, door to door, to many of the places in Tokyo we normally go. Because of the hill and the wasted commuting time, we eventually decided to move to someplace where we would have a flat walk to a more "strategically" located station.

After 16 months in the Tamagawa Gakuen, we moved to a 100-square-meter two-story detached house in the town of Koganei, about 25 kilometers west of central Tokyo and about 10 kilometers north of our previous apartment. It was a 6-minute, level walk to the station and a 30-minute local train ride to Shinjuku. The house also had three bedrooms (the middle-sized one had tatami and shoji, and the other two had bare wood floors and regular doors), a living room, a dining room (western style), a kitchen, a carport, and 1 $1/2$ baths. The kitchen had ample storage space, but the range had no oven and only three gas burners. It costed 260,000 yen per month for rent, which should be lower as of 2000, and we had only two major housing problems. First, we needed to buy rugs for the beautiful wood floors, and a dining room set. Second, we were very cold in the winter, because the stairwell, bathrooms, and hallway were unheated. It is standard in Japan to heat and air-condition only the main rooms, only when they are in use. Other than that we were quite content with the new place.

TRANSPORTATION

The Japanese train and subway system is outstanding, since it is extremely safe and efficient. Trains run strictly on schedule, and they deliver you close to just about anywhere you want to go. Many foreigners, including us, have no desire to own a car and are perfectly happy with a combination of public transportation, walking, and taxis. In fact, we do not miss driving and are considering repatriating, when we return to the United States, to an area where we will rely less on cars. Even so, traveling from one place to another has been the largest adjustment that we have had to make in Japan. Part of the reason that we were initially overwhelmed by traveling in Japan is that Tokyo is the first really large city that we have ever commuted to. Having oriented our lives around cars in small and medium-sized cities in the United States, we never dreamed how much time spent climbing stairs and walking would be required in a mass transit—oriented society. Because almost everything we do centers on central Tokyo, we still spend far too much time on the trains. We expected central Tokyo to be crowded, but our expectations have been exceeded. The intense crowding on the major train lines during peak times make it a chore to ride the train and walk near business-district train stations. But we never dreamed that the small-town train stations outside Tokyo would be so crowded. There are so many taxis, cars, bicycles, motorcycles, motor scooters, and pedestrians on the narrow streets around our new train station at peak hours that it is a safety hazard. To further complicate the traffic situation, there are no sidewalks along these streets. We doubt that people can be more aware of the crowding of Tokyo than when traveling.

SHOPPING

All of the necessities of life are available from the less expensive, second-tier stores. There is a domestic or licensed equivalent of virtually every western product, and so there are very few imported items available at most Japanese stores. And because of the small number of foreigners living in Japan, most imports are mainly suited to or modified for Japanese needs and tastes. This means that many foreigners, especially westerners, have trouble finding properly fitted clothes, their favorite foods, and many familiar housewares. Whenever we go overseas we bring back the maximum quantity of items that we miss from home but cannot find in Japan. In fact, on accompanied baggage, a surprising number of items carry little or no duty. However, as in all countries, the duties are inconsistent and it is best to look into the regulations before blindly bringing large quantities of anything. Part of our two-year supply of asthma medication was refused entry as "excessive quantities." Some of our friends brought in a case of champagne and close to a case of good liquor and paid about one U.S. dollar per bottle in duties.

LANGUAGE, PEOPLE AND CULTURE

Obviously, it is best to learn Japanese before moving to Japan. Predictably, very few foreigners have this skill when they move to Japan. Most Japanese have studied upward of half a dozen years of English from junior high through high school. This is a misleading statement, because most of this effort is spent preparing for the national and private college entrance examinations, which are written tests. Thus, relatively few Japanese are skillful at English conversation, even though they have some knowledge of English. Most white collar and engineers have a grasp of English, and all of our business is conducted in English. The second-tier stores where we do most of our shopping generally have no employees who speak English very well, and we usually end up using gestures, a few English expressions, and our very, very minimal Japanese.

The Japanese are hesitant yet curious when meeting a foreigner. There are very few westerners living in and around Tokyo and fewer still in the outlying areas of Japan. Except for those involved in international trade, very few Japanese are at all comfortable with foreigners. When dealing on a personal basis, they are polite and formal to the point where it gets in the way of trying to get acquainted. But in a crowd, especially on the platforms at train stations, all manners are suspended. The jostling can be unsettling, and no one hesitates to smoke in these crowds, even during the no smoking hours. (Since the latter half of 1990's, non-smoking cars, seats and places have become common in Japan and people more or less came to abide by the non-smoking rules.) But we have had Japanese strangers, both young and old, give us small presents in stores. There is a certain irony in that they do not know what to make of us and after two years we are still equally uncertain of them.

For our first six months in the old apartment, none of our neighbors paid any attention to us. After that, people in the neighborhood were greeting us and occasionally practicing their English on us. The constant invitations made so that people could really practice their English and become somewhat of a minor irritation after two years or so of living in Japan. But the shopkeepers began to greet us, and we started to feel that we belonged there. Since we moved to the new house, the people seem to be warming up to us a little more quickly.

CONCLUSIONS

It is our recommendation—and conversations with other foreigners lend further verification—that Japan should be regarded as an excellent place to pursue career and business interests. Additionally, it is possible, but takes considerable effort to live a reasonably comfortable family life reasonably close to the western standard of living. Also, do not move to Japan without a preliminary house hunting and reconnaissance trip. We would have shipped more of our belongings had we known what we know now. Living in Japan is not for everyone, but with the right attitude and sufficient motivation it can be done. We stayed here for a total of about five years from March 1989 to the mid 1990's.

From the Editor

Price and other situations from the late 1990's to year 2000 were added at the editor's discretion.

Hiroshi Honda

PART 4

Cultural Gaps and the Language Barrier

12

General Notes on Adapting to Life in Japan

Daniel K. Day

Living in a foreign country strains one's mental capacities in unexpected ways, and so I would urge anyone moving to Japan for a long period to read up on the country as much as possible before coming. I list some recommended readings at the end of this chapter; history and sociology are not everyone's cup of tea, nor are they mine, but knowledge of them in this instance is indispensable.

CULTURAL DIFFERENCES

I think the most helpful attributes for non-Japanese wanting to live in Japan are intuition and a sense of humor. I will concentrate in the following paragraphs on negative things, since the unpleasant surprises are what newcomers need most to be warned about. I hope the reader can believe that there will also be many pleasant surprises. I love Japan and feel as much at home there now as I can imagine feeling anywhere.

Intuition

Intuition will be invaluable for telling you what is going on when the words are not enough; this is important because the words can be deceptive if taken literally in Japan. As you will hear again and again, Japanese are concerned about maintaining harmony, or at least the appearance of harmony. This often results in people's not using the word for "no" even though that is what they mean. You must learn to read the signs of hesitancy and apparent waffling (and distinguish them from simple nervousness at dealing with a foreigner) which mean that a person is trying hard to think of an inoffensive way to refuse. A typical pattern of polite refusal found in the workplace is for someone to call over a senior, who in turn asks a coworker, who yet again asks his senior, who summons someone else, until you are surrounded by the makings of a committee, whose purpose, you have dimly realized, is not to tell you "yes." This is neither waffling or foot-dragging, it is protocol, and for you to bring it to an end before its time by blurting out an apology over your shoulder as you beat a hasty retreat would be unmannerly and counterproductive for the next time (you never know!) you go to the same office.

Haragei means the art *(gei)* of the belly *(hara)*, where belly signifies one's "heart," what one is really thinking. This art is in transmitting one's intention without putting it directly into words. The complimentary process is to read *(yomu)* another's intention, and that is called *hara o yomu*. Humans everywhere do these things, but the Japanese language has an awe-inspiring number of expressions for them. This reflects, I think, the importance accorded to intuition in everyday dealings in Japanese society.

And sometimes Japanese people "lie." The first time you "smell a rat"—when someone listens to your rendition of a Japanese sentence, looks utterly mystified, and then remarks that your Japanese is really good—may mark your initiation of what Japanese call *honne* and *tatemae*. A common American phrase is "maybe that's what he said,…" with the pause indicating that whatever it was "he" said should be taken with a heavy dose of salt. This corresponds to the Japanese tatemae, while one's true intention (hara, in the previous paragraph) is honne. It would be unjust to get angry if you are a victim of tatemae; chalk it up to experience and keep trying to read the signs.

Slow Decisions

The breakneck pace of the urban centers notwithstanding, it is a given that group decisions in Japan take time. I have never worked in a Japanese company and cannot recall any anecdotes to illustrate this, but the western top-down management system which allows—indeed, forces—fast decisions does not exist here. Instead there is (have you heard this before?) decisions by consensus, built painstakingly by those with the ideas by the process of persuading individuals to go along with their plan. The process is called *nemawashi*. Japanese are inured to these delays, even if they are personally irritated by them. You must adjust to them as well.

Reluctant Discussion

A westerner leading a group must adopt a different style when he or she wants people to contribute ideas. A common western pattern is for the leader to describe the plan of action and then to ask if there are any questions. This invitation will virtually always be greeted by a stony silence in Japan. This is partly because group discussions are not part of the Japanese educational process. It is best not expect members to speak out in groups at all; individually, though, people will feel more free to expose their ignorance ... or brilliance.

Another facet of this, by the way, is one of the hardest things for non-Japanese to accept: the refusal of most people to discuss politics or anything else on which there could possibly be serious disagreement. This dovetails with the Japanese concern with maintaining harmony, as you will hear time and again; still, it may well be one of the weakest points of Japan as a supposedly democratic political unit, as all too many seem to refuse to take any kind of political stand at all. To the westerner who wants to discuss political and social trends and expects some educational disagreement as a matter of course, it can make the people, ultimately, a little boring to deal with on a personal level. Given linguistic skill and some time of acquaintance, though, some Japanese will open up.

LEARNING JAPANESE

There are a wealth of materials for learning the language; if you go to school, your choice will already be made. If you plan to take lessons from a private teacher who is flexible about materials, I recommend the University of Hawaii Press series *Learn Japanese,* Volumes I through IV, for studying conversation. As for dictionaries, *romaji,* (Roman alphabet) dictionaries featuring alphabetic renderings of Japanese words with their English equivalents will be valuable at first, but are usually rather limited in vocabulary. Kenkyusha and Sanseido each put out handy English-Japanese/Japanese-English volumes for about 4000 yen. The

Japanese-English listings are given in *kana* (Japanese syllabic character) order; this will be bothersome at first, but will build your reading ability in kana and will drill you in the order, helpful once you develop the confidence to use directories, in buildings, for example. The English-Japanese listings will usually yield a line of kana and some frustrating kanji (Chinese characters), but sometimes you can simply show the printed page to the person to whom you are talking.

Do's and Don't's

I have taken few lessons in Japanese since coming to Japan, but judging from those and from my experience on the other side of the desk, I am confident about recommending some general do's and don't's. (1) You should set and keep in mind your goal in learning Japanese. How much do you want to learn, and how fast do you want to learn it? This is essential when making hard decisions whether a certain class or teacher is providing what you need. (A maxim: The object is communication in Japanese, not perfection in Japanese.) The ideal would be to observe a teacher at work before laying down any money, but I hear that most schools refuse to let prospective clients watch classes. Corollary to this is: (2) Do not allow schools to place you with students with a far different skill level. Your company may bring in a teacher and prefer to put everyone together to economize. Try to convince your manager that the company's money is best spent on lessons that are not a constant source of frustration to one student or another. (3) Small group lessons (two to five students) are the best; the individual gets plenty of attention, but has a chance to relax while other students interact with each other or the teacher. (4) Don't pay attention to people who tell you "they all speak English." In the first place, they don't. In the second place it is common courtesy to try to learn the language of your host country. In the third place, it will open up doors, here in Japan and everywhere else there are Japanese, who have an enormous respect and gratitude for people who make the effort to learn their tongue. In this, they are no different from many peoples of the world, with the sad exception of the Americans, who have a wondrously arrogant lack of appreciation of the efforts of foreigners to learn English. (This, I think, is one of our mental blocks in studying languages; we fear that all our work will go unappreciated.) (5) Don't believe people who tell you just to use the short forms of the verbs (you will understand what this means after you have studied for a few months) because "nobody uses the long forms." Oh, yes, they do. An explanation of these is outside the scope of this book; let it suffice to say that too much use of the short form is as irritating in Japanese as to much use of profanity is in English. (6) *Nihongo* is not the exclusive possession of the Japanese people. Whatever you learn is *yours,* so fine-tune your vocabulary for your own purposes. Make extra efforts. Look up words in your dictionary and learn vocabulary that is necessary and interesting to you. Try to express your sense of humor. (7) Sports are one of the very best ways of meeting Japanese in informal situations that do not involve drinking. Too often, conversations between Japanese and foreigners get onto one or the other of the languages themselves as the subject. Conversation, and your relationships with the Japanese people, become more meaningful when you meet people in situations where you are *doing* something.

My Principles in Studying Japanese

Learning and retaining vocabulary is the final challenge of any language; you will be working on vocabulary long after you have attained a command of the grammar, and so I would advise a frontal assault on this. Different studying techniques work for different people, and so more important than my ideas is to be creative in trying to find ways that will work for you. Try not to overdo any particular method.

Vocabulary.   The primary rule is to learn first to *recognize* the meaning of a word when you hear it (write it down for study in *hiragana,* the kana for native Japanese words, not in kanji), and second to learn to *say* the Japanese word after seeing the corresponding word in your own language. Do not be ashamed or discouraged by your need to review constantly. My method is to make lists and keep them in a standard-size notebook or small memo book, glancing at them at odd moments and muttering them under my breath. Two of my friends who dislike lists have other methods. One carries around 20 or 30 scraps of paper in his pocket with the English on one side of the scrap and the Japanese on the other. He pulls out several scraps at a time and goes through them; when he has learned the word, he throws the scrap away. The other friend uses scraps of paper all together in a "grab bag" which he reaches into at odd moments.

The Kana.   After you learn how to write the hiragana, use them. Learn and write out the names of your coworkers, of stores and restaurants. There is plenty of geographical information which you will find useful in daily life, and it can be combined with practice of hiragana. Write out the names of the subway stops, main streets, neighborhoods, wards, nearby towns, ski resorts, the 47 prefectures, and regions (Chubu and San'in *Chiho,* for example). Write out colors, tastes, shapes, textures, emotions. After you learn *katakana,* or kana for western-derived words, copy down the entire contents of the menu one morning as you sit in the coffee shop. Take home a paper menu from Kentucky Fried Chicken and copy it. Write down all of the animal names which should be written in katakana (*hamustaa,* for instance), the names of the elements, countries. This practice will also help you to remember how to pronounce words, which are Japanese words and should be pronounced in the Japanese manner, as I note later on.

The Kanji.   The first goal in learning kanji is to be able to read signs. You need numbers and prices (15 kanji are used for these; they're an easy place to start); next, the days of the week (8 kanji). After these, you need addresses: learn to write your own address, then the other wards in your city, main subway and JR (Japan Railway) stations, and nearby towns. Note casually what individual kanji mean, but concentrate on your goal, addresses. Always try to write the strokes in the correct stroke order; this will help you eventually when puzzling out someone else's scribbled handwriting. Signs in the train stations will begin to make sense. The visual noise on business cards will resolve into real information. People's last names, which are generally based on place names, will begin to look familiar. If your interest in kanji is piqued by this, invest in a copy of Gakken's excellent kanji dictionary-textbook and ask your teacher or friend to pick out a few vocabulary words under each character as useful as for you to learn. As you do this, give priority to learning the word in its spoken form. This sounds contradictory, but if you practice learning to say the word while writing it, you will surely learn to read it. A word you can say and read but not write is worth more than a word you can read and write but not pronounce. Try to learn at a leisurely, steady pace, say, two kanji (six to eight vocabulary words) per day, five days a week. Review constantly. Kanji must be *written* (not just stared at) to be remembered, while vocabulary must be *vocalized.*

Pronunciation.   Again the maxim: Your object is to learn to communicate, not to be perfect. That said, it is worth pointing out some pitfalls. Recent language-learning theory holds that there is a psychological "monitor" which operates while one is learning and speaking a foreign language and spots mistakes, or, on the other hand, refuses to spot mistakes. Adults tend to feel ridiculous imitating sounds which do not exist in their own language: the monitor malfunctions. The self-consciousness, if you suffer from it, is a delusion and should be fought with determination. Second, modern Japanese has many katakana words which were borrowed from English. These must be pronounced in the Japanese fashion, not English. Imagine someone saying "hors d'oeuvres" in its French pronunciation in the midst of an otherwise English sentence, and you can probably understand how comical it sounds to mix pronunciations.

SUMMARY

There is no substitute for being able to speak Japanese. The writing system and the completely different grammar make for a formidable challenge, but they can be mastered with perseverance. I firmly believe that anyone who makes this effort will never forget it.

References

I recommended these books (and one article) on Japan, in order of priority:

Edwin O. Reischauer, *The Japanese*, Harvard Press, Cambridge Mass., 1977.

Boye De Mente, *Japanese Etiquette and Ethics in Business*, 5th rev. ed., National Textbook Company/Passport Books, Lincolnwood, Ill., 1986.

Frank Gibney, *Japan: The Fragile Superpower*, NAL Books, New York, 1989.

Karel van Wolferen, *The Enigma of Japanese Power*, Knopf, New York, 1989 (also Random House/Vintage, New York, 1989)

Akio Morita, *Made in Japan*, New American Library/Signet, New York, 1988.

Richard Mason and John Caiger, *A History of Japan*, Fress Press, New York, 1973

Linda Sherman, "Breaking the Intimacy Barrier," Japan Quarterly (Asahi Shimbun), July–September 1990.

Textbooks for the Japanese language:

John Young and Kimiko Nakajima-Okano, *Learn Japanese: New College Text*, Vols. I–IV, University of Hawaii Press, Honolulu, 1984–85

Kuratani et al., *A New Dictionary of Kanji Usage*, Gakken, Tokyo.

13

The Cultural Gap Experienced by a "Gaijin" Engineering Executive in Japan

Raymond C. Vonderau

INTRODUCTION

In this presentation I will discuss my experiences and observations as a western-trained engineering executive in Japan that illustrate differences in practices in the engineering profession between Japanese and western society. The sum of these differences as they affect any one individual constitutes what is referred to in the title of this chapter as the "cultural gap."

The foreign-trained engineer can minimize the effect of the cultural gap by knowing about these differences in social culture (personal life) and corporate culture (working life). Through knowledge and understanding of each other's cultures, progress toward a *global* engineering community can be achieved. As the need for more foreign-trained engineers increases, it is hoped that this knowledge and understanding will help us to produce a group of professionals who can be considered *international* engineers rather than foreign engineers.

In identifying these differences, it is not implied that the Japanese or the western way is wrong; however, these differences can cause difficulties for the foreign engineer which may keep him or her from performing as anticipated and from enjoying the opportunities provided by the new environment.

In discussing the cultural gap with both western-trained and Japanese engineers, I found that it had many different associations for different people. Since the experiences making up the cultural gap are very personal, each person is affected in a different way depending on his or her personality, interests, and particular situation. By receiving input from others, I hoped to present a broader range of experiences and opinions than only my own. Therefore I am indeed grateful to those people who shared their experiences and opinions with me in confidence.

My experience in Japan is unique, as is each foreign engineer's experience. I work alone as my company's representative in the offices of a major Japanese corporation which has been a licensee of my company in the United States since the 1950s. I am the ninth person from my company to be located in this assignment. Therefore my daily experiences are the result of a mature intercultural relationship between our companies. My experiences could thus be quite different from those of an engineer who is the first foreign person to work in his or her Japanese company.

Most Foreign Engineers' Assignments Are Temporary

There seem to be relatively few foreign engineers in Japan. The ASME Japan Section has about 700 members as of 2000, of which only about 20 members are foreign engineers, and the JSME has about 45,000 members, of which fewer than 100 members are foreign engineers. The assignments of these foreign engineers are somewhat varied, but a majority of them seem to be in academic and research fields in Japan. These assignments appear to be temporary in nature compared with a line assignment in the engineering department of a Japanese corporation. This is interesting because it seems to reflect the difficulty of making a permanent commitment to the "corporate fraternity" of a Japanese corporation, from which Japanese engineers have expectations of lifelong employment and regular advancement in job position along with their peers.

Hiring Practices

If one understands the Japanese educational system, the practice of hiring only graduating engineers into the company (who traditionally begin their employment on April 1 as freshman) becomes quite clear; and if one understands the strong ties that develop between individuals and their company from this practice, then the difficulty of including foreign engineers in this system, as it exists, seems inevitable.

The larger Japanese corporations will also solicit and hire experienced engineers in order to have their expertise available for specific needs or projects.

Joint ventures between Mazda and Ford and between Mitsubishi and Boeing are typical examples of such projects. Even though the scope of these projects may extend over a period of several years, this still falls short of a lifelong commitment to the corporation and for all practical purposes is a temporary assignment. I would conclude that, at this time and for many reasons, it is unlikely that many foreign engineers will make the lifelong commitment to a Japanese company that is necessary for inclusion in the Japanese "fraternity," with its benefit of job advancement.

Therefore, it seems that foreign engineers and Japanese engineers must learn to work effectively together even though the foreign engineers are not included in this fraternity. For a foreign engineer to be comfortable and effective in his or her job position and assignments, it is important that he or she understand the Japanese engineer's relationship to other members of the company and particularly how the Japanese system works. There is no written code of ethic that dictates the expected behavior from this relationship; such a code is not normally needed when practices stem from cultural roots. The foreign engineer, then, is obliged to learn through observation and perception about the pressures that mold corporate behavior.

THE JAPANESE COMPANY ENVIRONMENT

When observing Japanese engineers to learn the Japanese way of working life, the foreign engineer sees many activities, relationships, and peer pressures that differ from those experienced in western society. The most obvious difference, the first to be seen, is the arrangement of the office. In a Japanese office all the desks are arranged close together in clusters in a large room with relatively few or no separate enclosures. This makes working conditions very crowded, noisy, and disruptive of concentration. The large room appears to be difficult to heat uniformly in the winter, since it is usually too cold for comfort. Similarly, in the summer the room is usually too hot for comfort. Not every desk has a telephone, and to make or receive a phone call it is necessary for most of the engineers to leave their desks and use a phone at a centrally located desk. The foreign engineer can adjust to these differences, but until that happens, his or her contribution may be adversely affected.

Meetings

Many meetings are held during the working day to deal with any number of items of general company business: for example, weekly safety meetings, earthquake or fire drills, meetings for issuing routine announcements or planning company social projects, and periodic meetings to review the status of projects. The western engineer will soon note that there is very little participation by the individuals attending these meetings. This is especially surprising with regard to the project review meetings, since westerners are accustomed to using such meetings as an opportunity to contribute to the project. In Japan, however, the project group leader or manager does most of the talking during these meetings. Seldom is any criticism introduced, and seldom are alternative suggestions voiced which would have an impact on the project objectives or change the scope of study. This makes it difficult to know the true feelings of the Japanese engineers involved with the project.

Importance of Group Acceptance

To the foreign engineer, there appears to be a little opportunity for individual creativity, since the total group has the responsibility for its members' contributions to their assignments and the project activities rather than the individuals themselves. Members of the group tend to become specialists in a relatively few disciplines. It also appears that members of the group are never aggressive but always conform to the expectations of the company, managers, and peers. It is obvious that acceptance and support from the group are very important to the Japanese engineer. Nothing is done to jeopardize this relationship. Therefore the working day is long—usually 10 to 12 hours, in order to strengthen the bonds holding the group together, even though not all that time is necessarily spent productively.

Vacations and Personal Time

Members of the group seldom take their allotted vacation time; one of the primary reasons seems to be the need to show support for the group's objectives. Japanese engineers are rarely away from the office except when they are traveling on company business or when the office is closed on Sundays and holidays. As a result of this peer pressure, the foreign engineer will notice that married Japanese engineers have little opportunity to spend time with their families, except perhaps on Sundays. These demands on a husband's time (as is usually the case) require strong support from his wife to run the household and to provide the necessary parental guidance for their children. In contrast, a foreign engineer is expected by his or her company to take any allotted vacation time.

PLANNING AND FOLLOWING A CAREER PATH

Several times in the past few years, I have asked young Japanese engineers what their career goals were and whether they had a plan or timetable to accomplish their goals. Nearly always the response has been that they have no personal goals or ambitions but rather only wish to be successful within their company by doing as well as possible in each of their assignments. Japanese engineers seem to be very willing to leave their career path planning totally in the hands of the company even though the company's plans are not made obvious to them. In contrast, the western-trained engineer is encouraged to set personal career goals and only to solicit the company's support to accomplish them.

The practice of annually promoting or moving people, particularly in large Japanese corporations, offers some opportunities for changing job assignments; however, many times changes are made without consulting the individuals affected by them. This can create problems for the individual and sometimes results in separation from the family when the new assignment is in another city or in a foreign country. It seems to be the case that employees do not refuse a company-planned career move, because doing so can seriously jeopardize opportunities to be considered for future assignment changes or promotion within the company.

On the positive side, the frequent movement of people within the company, coupled with lifelong employment, provides an excellent way to train company-oriented and knowledgeable managers. Strong managers who can lead people to accomplish the company's goals are one of the very positive results of the Japanese way of career planning, which allows the strongest person to achieve the position of leader of the group.

SOME DIFFERENCES THAT CAN CAUSE PROBLEMS

At the same time, there are some differences in the customs that do continually cause difficulty for the foreign engineers in the Japanese work environment.

One of these differences is that to be successful in a Japanese company, a foreign engineer must be able to work alone. This is partly because of the language barrier and partly because of the work ethic of the Japanese. There appears to be little fraternizing or sharing of knowledge between peers even within the same group. Knowledge is shared more readily with subordinates, on an "as needed" basis, but with peers it is rarely shared even on an informal basis.

Demands on Employees' time

In most Japanese companies, there are many traditional company-sponsored activities that are essentially social activities, since they are held outside the company. These are often ambitious programs, and the participation of each group member within the department is expected. Because the programs are ambitious, time for planning and preparation is required, which is normally taken out of busy working hours. These activities enable the employees to know each other better, and when the events are competitive, esprit de corps and camaraderie can be fostered; this is expected to enable the employees to work together in better harmony. By assigning different leaders or planners each year for these traditional events, some insight into each person's interest in and ability for leadership can be obtained. This can be helpful in identifying the "natural" leaders when the time for promotions arrives.

The total commitment of time–not only the lengthy workday, but also these additional company functions–is very difficult for the foreign engineer. Participation in these events, in addition to their working hours, makes it difficult for foreign engineers to find time to pursue any hobbies or activities of personal interest which are not related to the company or its activities. Therefore, foreign engineers may be concerned about becoming one-dimensional persons. The freedom of choice enjoyed in the home country is not available; one cannot pursue activities for personal growth and satisfaction, because the time is not available. The total commitment of time for the company's needs and objectives is difficult to accept for the foreign-trained engineer.

Differences Outside the Workplace

The foreign engineer can normally learn about and accept the Japanese customs in the workplace that have been described in this chapter. However, in order to fully enjoy living in a fascinating country with people of a different culture, it is also necessary to accept and adapt to many differences of a personal nature. As previously stated, the foreign engineer's stay in Japan is usually for a relatively short time. For this reason it may actually be easier to adjust to the kinds of differences in everyday living conditions that guidebooks often refer to as "culture shock." Some important concerns are as follows:

- The very crowded conditions and close proximity to people encountered in living, working, commuting, shopping, and enjoying recreation will cause mental fatigue in the foreigner first coming to Japan.
- The cost of rent, food, restaurant meals and all Japanese goods and services was very high especially in the early to mid-1990s (but has become much lower as of the year 2000); however, it is also obvious that the quality of Japanese goods and services is very high.
- In most cities it is not possible to conveniently drive a car to accomplish personal errands, and the newcomer should accept the fact that ownership of a car is not always practical or necessary.

There are also many differences of a personal nature which make Japan very attractive to the foreign person:

- The politeness, courtesy, and work ethic of the Japanese people
- The many closely knit families with happy, well-behaved children
- The safety one feels in any part of the cities
- The efficiency of the transportation system
- The cleanliness of the Japanese people and their cities and living environment
- A strong respect for nature and a desire to preserve the beauty of the land, mountains, and sea
- A fascinating culture that teaches self-discipline and respect for others

In addition to the professional benefit of working in a Japanese company, all these are things a foreign engineer can enjoy by making the effort to bridge the cultural gap.

THE FUTURE: NARROWING THE CULTURAL GAP

If the shortage of technically trained professional people in Japan continues to become more acute, more foreign engineers are likely to be needed by Japanese companies. As suggested previously, it is likely that the foreigners hired will continue to be experienced engineers who are selected for having particular knowledge useful for a specific project need.

Considerable improvement in the necessary cultural adjustment is possible if these candidates can be taught–before they come to Japan and again when they arrive in Japan–about the known differences between Japanese and western society that will be encountered in living and working. Besides shortening the personal adjustment period, better knowledge of cultural differences would enable the newcomer to make an almost immediate contribution to the group's project objectives.

Better knowledge also means an understanding of the cultural differences on the part of Japanese engineers and managers. Their interest could minimize the effect of these differences by leading management to make just enough changes in the traditional company system to facilitate the integration of the foreign engineer into the Japanese company family, even though he or she may have no intention of joining the Japanese "fraternity" of lifetime employees. One Japanese "sponsor" family could be appointed to assist and monitor each foreign family's acceptance and understanding of this new social culture; interested company members could volunteer. In addition, a company or a corporate sponsor could be appointed to assist and monitor each foreign engineer's understanding of the corporate culture and the Japanese way.

The advantages to the Japanese company and to the foreign engineer in narrowing and bridging the cultural gap will be a more productive, more company-oriented, and totally involved *international* engineer. I think that this approach might provide a blueprint for a technical community that works together *on a global* basis—across national boundaries.

14

Be Wary of Generalizations, Be Humble, and Be Fearless of the Language

Craig Van Degrift

Understanding Japan is like examining an onion: one layer after another is probed only to reveal a deeper layer inside. The visitor passing through hastily on a business trip sees a western-style country that is peppered with shrines, temples, and castles which hint of a mysterious past. The Japanese seem to match their reputation for being a homogeneous swarm of tireless workers single-mindedly dedicated to their jobs. Even though their language is said to be the most difficult in the world, their literacy rate is the highest.

Visitors staying for a month will have explored a little deeper and gained some insights missed by the one-week visitor. If they stay for three months, they finally realize that understanding Japan is much more difficult than they had thought. After a year, their Japanese language study is likely to have made significant progress and they are likely to have made Japanese friends with whom they can have detailed discussions of political and cultural questions (probably still in English). If their friends do not represent too narrow a slice of Japanese society, they can finally start to really understand Japan.

The different writers' contributions in this book should help speed up this process, but the reader must be careful to remember that Japanese society is *not* perfectly homogeneous. Different observers' views will be shaped by those individuals with whom they interact as well as by their own individuality. *One must be wary that the unavoidable generalizations used to describe the traits of Japanese society as a whole not be blindly applied to individual people or a specific workplace.*

The views I present in this chapter are those of an American government scientist who worked for a year in Tsukuba at the Electrotechnical Laboratory (ETL) of the MITI (Ministry of International Trade and Industry), a particularly western-orientated laboratory in a newly created "science city." Prior to my arrival, I already had a number of close friends among Tsukuba's scientists, because my home laboratory, the National Bureau of Standards (now called the National Institute of Standards and Technology), regularly hosted Tsukuba scientists. My interest in the Japanese language had started around 1960 while I was in high school in Los Angeles, although it lay dormant during most of the interim.

I found the Electrotechnical Laboratory to be very similar to my home laboratory with respect to quality of equipment, facilities, and personnel. While individual offices were rare, everyone had his or her own telephone and adequate storage space. The group offices were relatively quiet and comfortable (our standards work required air conditioning), although there was a problem with "secondary" cigarette smoke.

The Japanese often speak of their nation's lack of creativity, but this was not evident at ETL. Even though its scientists and engineers have been processed by Japan's strict school system, the particular atmosphere of ETL allows a rekindling of their temporarily suppressed innate creativity. (The unusual nature of ETL may in part be due to the fact that a rather large percentage of its scientific staff has spent a year or more at research laboratories in Europe or America.)

Similarly, the reputed homogeneity of Japanese society does not exist at ETL. There seemed to be no less individuality among ETL scientists and their families than among their counterparts in America—some worked 8 hours a day, others 14; few wore suits; some fathers helped change diapers at home, others didn't; some repaired their own cars (one owned a 1964 Volkswagen beetle); and the scientists and engineers included some women together with many men. One difference, however, was that the ETL scientists seemed to read more technical literature (largely in English) and to attend fewer conferences than their contemporaries in America: there were many informal study groups that gathered to work their way through books (usually English-language books) in order to gain expertise in new fields. The most remarkable difference between ETL and a similar American organization, however, was in the laboratory's management. The characteristically Japanese bottom-up decision making was apparent, as was the informal bypassing of some formal bureaucratic barriers.

Surely, the work environment at ETL is exceptional for Japan, but that it exists at all at a major government laboratory underscores the danger of trusting too heavily in the usual generalizations about Japanese society. Japanese society and school traditions do *tend* to suppress individualism and creativity, but they by no means eliminate them.

While social interaction involving alcohol is common, I found that my absolute abstinence from alcohol did not create undue difficulty. Nonconformity of a foreigner, if it is not flaunted, is tolerated, and perhaps even admired.

It is important to be humble. It is very easy to be wrong anywhere, but when you are wrong in Japan you are unlikely to be clearly corrected. In fact, strong expression of opinion is likely to create serious barriers to further communication. Never underestimate your unassertive Japanese colleagues. Try to convey an uncertainty about the opinions you express. Using "I wonder if...," "Could it be that...," "Is it possible that...," and similar expressions can greatly soften one's speech. This tradition may be changing: one of my Japanese friends (observing my children) seriously asked, "How do American schools teach children to argue?" He was genuinely concerned that Japanese children should learn how to have more vigorous discussions.

It is easier to be *genuinely* humble in your ideas if you think more deeply about a given topic. For example, if you find yourself thinking, "The Japanese really need more big discount stores," remember that low prices and wide variety are not the only factors to consider. Walking to the corner store is something that your kids can do unsupervised, is healthy exercise, doesn't use gas, and allows you to get by with a more compact refrigerator if you have a small apartment. Also, in the United States as well as in Japan, the mom-and-pop stores employ a segment of society which might otherwise be unemployed or on welfare. Questions that seem to have obvious answers in American society can turn out to be more complex when fully analyzed. Japanese consideration of the long-term effect of procurement and sales decisions may lead to business practices which seem peculiar to foreigners.

Part of being humble is being willing to physically help in preparations for an activity as well as the cleaning up afterwards. Custodial help does not exist in schools and is minimal in workplaces. Always be on the alert for ways in which you should be helping in maintenance and other group activities, both at your workplace and in your neighborhood. The cleaning of trash pickup sites in residential areas and certain school maintenance tasks that the children cannot do are performed by residents and parents according to an established schedule.

Many Japanese scientists and engineers can understand English if one uses clearly pronounced, idiom-free English supported by clear diagrams. Nevertheless, a modest command of Japanese will allow you to better communicate with the people in your firm who are likely to be non-English speakers—the machinists, production workers, and secretaries, for instance—as well as allow you to handle your personal life independently. If each interaction with lower-level workers requires the help of a Japanese colleague, your value to the firm will be diminished.

Even though full fluency in Japanese requires more than five years of intense study, every effort spent learning Japanese will yield benefits. It is quite a shock to well-educated visitors to realize that they are virtually illiterate! Simply attempting to study Japanese is an important social courtesy. It indicates an interest in the culture, an effort at self-sufficiency, and a recognition that communications with non-English speaking Japanese are valued.

One can first concentrate on learning useful subsets of the full language. By intensively using flash cards, the 46-character *katakana* syllabary used for imported foreign words can be quickly learned. Since there are more than 30,000 imported words written in katakana, this provides immediate help in understanding many signs and package labels and even in discerning the subject of the written text. Some skill and imagination, however, are required to correlate the Japanese and English versions of these words. Furthermore, they *must* be pronounced in the Japanese manner to be understood!

Next, one must learn the other syllabary (*hiragana*), start building a core vocabulary of genuine Japanese words, learn survival phrases and basic grammar, and begin learning the Chinese characters (called *kanji*). The hiragana requires the same degree of effort as the katakana, but provides only light immediate benefit until the core vocabulary is learned. Even though Japanese grammar has many levels of politeness and usage subtleties, learning enough for the purposes of basic communication is rather easy. Also, it is easy to learn to pronounce Japanese understandably. Be sure to get the vowel sounds right (there are only five), but ignore the tonal differences some books and courses try to teach.

It is the memorization of the genuine Japanese vocabulary and kanji that makes Japanese difficult. The words have no connection at all with those of western languages, and because they are built up from a small number of basic sounds, they are especially difficult to remember and to discern aurally. You should consult a small conversation textbook for a suitable vocabulary list rather than try to find Japanese equivalents to your favorite English words. Whereas children seem to be able to rather quickly learn to pick out words they know from rapidly spoken monotonic Japanese, this skill is only slowly learned by adults. Progress is greatly hastened if one dives in, not worrying about mistakes. Social interactions with non-English-speaking Japanese at singing bars, in athletic activities, or in craft clubs, for example, are very effective.

Kanji present the greatest linguistic barrier to foreigners. Fortunately since few Japanese words will be misunderstood if written in hiragana instead of kanji, and since Japanese word processors have become widespread, *you can postpone or even skip learning to write kanji.* Just learn how to count brush-strokes (to aid in using dictionaries), and concentrate on reading. Learning the meanings of the first 100 or so characters is easily accomplished using flash cards and a good pictorial imagination. Knowledge of these first few kanji can greatly help you in navigating Japan's transportation system. Learning the pronunciation, however, is not so easy, because each character will usually have two or three pronunciations depending on associated characters or text. The most serious hurdle to learning kanji, however, is encountered when spanning the gap between the first 200 kanji and the 600 or 700 level. In this range, it seems that characters can be forgotten as quickly as new ones are learned. Also, there are many characters which are easily confused, defy pictorial interpretation, or have ill-defined meanings when studied individually. Fortunately, as your capability approaches the 1000 mark, the characters once again become *easier* to remember. This happens because kanji have many subtle meaning and pronunciation interrelationships which are not apparent until a large number of base characters and character components are known. Hang in there!

References

Perhaps the widest selection of Japanese-language learning materials may be found at Bonjinsha Co., Ltd., in Tokyo or Osaka at the following addresses:

Tokyo: Second floor, Kojimachi New Yahiko Building, 6-2 Kojimachi, Chiyoda-ku, Tokyo 102. Above a Dairy Queen (opposite a Sakura Bank office), about half a kilometer east of the Yotsuya subway and train station on Shinjuku Street. Phone 03-3239-8673. Fax: 03-3238-9125.

Osaka: Second Floor, Iwamoto Building, 7-1-29 Nishinakajima, Yodogawa-ku, Osaka 532. On the second floor of a building between the Shin-Osaka train station and the Chisan Hotel. Phone: 06-6390-8461. Fax:06-6303-7049.

The following are listings for recommended books

An excellent, though expensive, set of pocket books written to help foreign visitors with Japanese culture, geography, and language is the illustrated book series published by the Japan Travel Bureau, Inc. volumes one through 13 (volumes 11 and 12 are in French). The first volume is *A Look into Japan,* Books Nippon, 1986 (ISBN 4-53300-30-79).

The best technical Japanese textbook is *Basic Technical Japanese,* by E. E. Daub, R. B. Bird, and N. Inoue, published for Japan and Oceania by the University of Tokyo Press (ISBN 4-13-087051-3;0-86008-467-1) and everywhere else by the University of Wisconsin Press, Madison, 1990 (ISBN 0-299-12730-3).

The best kanji-to-English dictionary is the *Japanese Character Dictionary with Compound Lookup via Any Kanji,* by M. Spahn and W. Hadamitzky, published by Nichigai Associates, Tokyo (ISBN 4-8169-0828-5 C0581).

There are numerous textbooks for learning general Japanese and books to aid in Kanji study. Of these, I recommend *Japanese for Today,* published by Gakken, Tokyo (ISBN 4-05-050154-6 C0081), and Kuratani et al., *A New Dictionary of Kanji Usage,* also published by Gakken (ISBN 4-05-051805-8 C3581).

The most complete Japanese-English and English-Japanese dictionaries are designed for use by Japanese and therefore assume complete knowledge of kanji. The beginner must select from one of the less complete, "Romanized" dictionaries designed for foreigners.

15

Bridging Gaps by Cultural Simulation

Kazuo Takaiwa

CULTURAL GAPS

Culture Shock and Cultural Friction

When people come into contact with natives of a different culture, they may feel uneasy and sometimes embarrassed. We call the uneasiness *culture shock;* some of the victims of culture shock suffer great anxiety and have to return to their home countries. Simultaneously, the newcomers may create a similar shock in their counterparts. A crunch may come: cultural friction. Friction may give rise to a sense of incompatibility, dislike, and, at worst, animosity. Once distrust has reared its head, it is time-consuming and expensive to wipe it out.

P. R. Harris and R. T. Moran report in their book *Managing Cultural Differences*[1] that "adjustment problems of Americans abroad are severe, and adjustment failures are costly in terms of economics, efficiency of operations, intercultural relations, and personal satisfaction with duty abroad." Moran and Harris reported that an American company dispatched about 300 engineers and their families to Iran and had to return 33 percent to the United States. Each family that had to be replaced cost the company $210,000 (p. 164.).

[1]Philip R. Harris and Robert T. Moran, *Managing Cultural Differences,* 1st ed., 3rd printing. Copyright © 1983 by Gulf Publishing Company, Houston, Texas. Used with permission. All rights reserved.

"Some anthropologists, such as Kalervo Oberg, consider culture shock a 'malady, an occupational disease which may be experienced by people who are suddenly transported abroad'" (Harris and Moran, p. 92). But "culture shock is neither good nor bad, necessary nor unnecessary. It is a reality that many people face when in strange and unexpected situations." Some of these people, if it is important enough to them, will nevertheless choose to make the most of the experience of another culture in spite of the possible dysfunctional effects of culture shock (p. 93).

Culture and Civilization

Ryotaro Shiba, one of the most famous novelists in Japan, discussed the differences between culture and civilization in an article that I would like to paraphrase here.[2] Civilization is rational, he said, but culture is nonrational. For instance, a red traffic signal tells all vehicles to stop and a green signal tells them to go ahead. This convention is used throughout the world. In the sense that traffic signals are rational and universal, they belong to civilization.

On the other hand, culture comprises conventions that are not utilitarian in purpose and are observed only by specific groups—for example, specific nations—and is therefore nonrational. One group's culture is not applicable to other groups. For instance, in Japan, it is deemed good manners when opening a *fusuma,* a Japanese paper door, especially when greeting guests, for the lady of the house to drop to her knees, place her right-hand fingers on the flat metal pull, place her left-hand fingers on the lower edge of the fusuma, and slide it back with both hands. Of course, she could open a door with one hand while standing on her feet, but this happens not to be considered good manners. This is culture. Nonrational rules are themselves part of the joy of culture and part of what gives it its meaning.

I think that this is a very reasonable summary of the subject.

Struggles between Cultures

There is no such thing as superiority or inferiority among cultures; there are only differences. For example, when Japanese people want to call a friend over, they stretch out an arm with the palm down and move the fingers, while

---

[2]Ryotaro Shiba, *Yomiuri Newspaper,* April 20, 1985.

Europeans and Americans perform the gesture with their fingers up. One is not better than the other, just different. Whenever a nation becomes overly proud and proclaims the superiority—rather than simply the uniqueness—of its culture, it seals its fate to be overthrown, no matter how great it may have become. Culture springs from the history and life of a nation in the past and is not enhanced by chauvinism.

Unfortunately, it happens all too often that arguments and rifts over superficial differences occur between partners in intercultural efforts. To promote understanding, the differences between Japanese culture and American culture were the theme during the recent Structural Impediments Initiatives in 1990 attended by the United States and Japan, which had been called by the American government to address the issue of trade imbalances. (One major difference is that democracy does not mean the same thing in Japan as in the United States.)

One and Fifty

The ratio of the population of Japan (126 million) to the population of the world (6 billion) as of 1999 is approximately 1 out of 50. "One and fifty" is my name for my concept of the proper attitude that Japanese should show during contacts with the rest of the world, as explained in my book *Abu Dhabi de Kaita Seiko suru Kaigai Bijinesu* ("success in business overseas: notes from Abu Dhabi," in Japanese). The book reports my key experiences in 15 years of assignments in Singapore, the Middle East, Africa, Indonesia, and South America, where I dealt with cultural friction between Japanese (including myself), the local people, and third nationalities. The projects involved several hundred or sometimes over 1000 Japanese engineers and supervisors, and several thousand local personnel. My philosophy is that since the Japanese constitute only one-fiftieth of the world's population, it is essential for us to adapt to the ways of the west and host countries once we have left our own country. Conversely, when in Japan to do business, it is reasonable for westerners to be expected to adapt to Japanese ways in business.

Local people in many parts of the world have become accustomed to international ways, that is, western ways of doing business, during the long colonial period. On the contrary, Japanese until recently had few experiences of trade with westerners because of their national isolation for 300 years. Japanese do not live in a society of contracts and job descriptions, but in a society of oaths. They are expected to behave on their own as good citizens. The small-group activities TQC and TPM (see Chapter 7) are not compulsory but are undertaken autonomously to promote productivity. Japanese social rules are quite different from western counterparts. That is part of the concept of "one and fifty."

Island Tribes and Continental Tribes

The Japanese are an "island tribe" (to paraphrase the Japanese term) that has developed its culture in relative isolation from other nations for more than 2000 years. Though peaceful intercourse with China and Korea has been recorded since the beginning of the Christian era, foreign commerce was strictly controlled by the imperial government once it had established itself in the sixth century. Japan has experienced only a few invasions by other nations. The first known in history were the Mongolian invasions of 1274 and 1281. On November 19, 1274, the invading Mongolian forces were devastated by a typhoon, which sank many of their ships as they began their retreat by sea. The next attempt, in 1281, suffered a similar fate, losing over half the force of 100,000 under the swords of the defending Japanese troops and a huge typhoon. This typhoon was called *kamikaze,* or "the wind of the gods," by Japanese historians. (From this experience, the word *kamikaze* became a metaphor for "luck" in Japanese. It was that desire for luck that led to its use in World War II—called the "Pacific War" in Japan—which ended when two atomic bombs were dropped on the country.)

Invasions of other countries by Japan have also taken place at only a few times in history. History tells us that the first was when Hideyoshi Toyotomi, the chief minister of state, sent troops to southern Korea between 1592 and 1598. The second was an invasion of Korea during the Sino-Japanese War, 1894–95. The third foreign campaign was in the Russo-Japanese War, 1904–05; the fourth, the attack on Qingtao (Tsingtao) in 1914. The fifth was the invasion of Manchuria and the Chinese mainland in 1931–40. The last was during the Pacific War from 1941 to 1945, at the end of which 4.5 million members of the Japanese armed forces returned to their burned-out homeland from the overseas dominions they had held.

Most civilizations other than Japan's have developed on continental mainlands, where borders provided by nature are less definitive. The peoples of such nations are called "continental tribes" in Japanese; these civilizations traditionally depended on territory to feed their populations and sought to increase their territory when their populations rose. (Starting with the Edo period after since 1603, food shortages and population increases did not tend to be a cause of war within Japan. The food supply came under control of the landlords, who encouraged development of farmland to prepare against famine.) Thus, continental tribes were constantly preparing to wage war, or to protect themselves against invasion. There was antagonism of one people against another; nations could not do business with each other without negotiations and contracts.

Wa (Harmony) and Antagonism

The relations typical of "continental tribes" were not an important element in Japanese domestic history. The Japanese, as a relatively homogeneous population, were not in a position to invade neighboring tribes or to be invaded by foreign tribes. Thus, Japanese had to live and stay in the islands in harmony with each other. To break harmony was tribal suicide.

The maintenance of harmony in social life, called *wa*, was required by the first article of the constitution of Shotoku Taishi (574–622), the most famous prince and one of the greatest men of wisdom in Japanese history. The concern with wa is echoed in the Japanese proverb "silence is golden but eloquence is only silver."

Japanese sought to be a single culture and to use a single language in their territory to keep harmony. If members of a different cultural group arrived on the scene, the Japanese tried to assimilate them so as to erase the sense of incompatibility. They lionized the newcomers at first, but then turned on them and forced them to acculturate. The histories of Europe, Asia, and Africa, on the other hand, are histories of antagonism, opposition, and self-assertion, as described by Paul Kennedy in his famous book *The Rise and Fall of the Great Powers*.[3]

CULTURAL SIMULATION

Principles and Procedures

I have developed a computer program of cultural simulation for testing and training in the ability to adapt to a foreign culture. (I have applied for a patent from both the Japanese and American patent bureaus.)

---

[3]Paul Kennedy, *The Rise and Fall of the Great Powers: Economic Change and Military Conflict from 1500 to 2000*, Random House/Vintage, New York, 1989.

On the screen there are still pictures and moving pictures of scenes you might see in the country of interest that might seem odd to an outsider. The pictures are followed by five sentences that make up a multiple-choice question about the scene. One choice is the reaction a native would have; it is the "right" answer. The others are graded by an expert system to show the test subject's "adaptability to the different culture (ATDC)." Thirty scenes make up one test, and their content depends on the target country, the occupation of the test subject, and the subject's past experience.

Depending on the subject's answers, the ATDC grade is at one of four levels:

1. Is ready to adapt to the different culture.
2. Has adaptability to the different culture, but needs some training.
3. Has adaptability to the different culture, but needs a lot of training.
4. Is not ready to adapt to the different culture.

In the cases that follow, the target country is Japan, the occupation is engineer, and the grade of the subject's experience is "elementary," meaning that the subject has never been in Japan or is in Japan for the first time now. Three examples are shown.

Example 1: CS-4, NOODLE. (See Figures 15.1 and 15.2). table manners differ extensively among peoples. Westerners use knives and forks; Japanese, chopsticks; and Indians, fingers. The custom for drinking coffee and especially for taking soup varies substantially; the Japanese technique might seem particularly strange to a westerner and thus is a good basis for a test question. In the case of eating noodles, a respondent would be rated as follows:

▶ Answer 1: SWALLOW NOODLES WITHOUT CHEWING. If the respondent thinks that swallowing noodles without chewing is good manners, he or she has little understanding of Japanese manners, but may be able to achieve a high ATDC after much training. Therefore, the respondent is given a grade of 3.

▶ Answer 2: TAKING UP A BOWL AND PINCHING NOODLES IS BAD MANNERS. A bowl of noodles is heavy. In Japan, taking up the bowl would not be good manners. That would be more in keeping with Korean customs. The respondent will have good adaptability after moderate training and is graded 2.

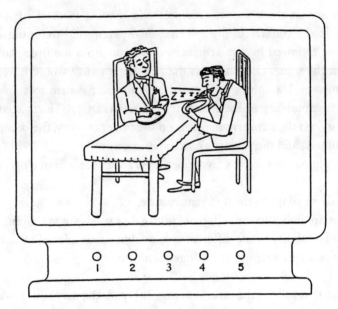

Figure 15.1   Scene: eating noodles.

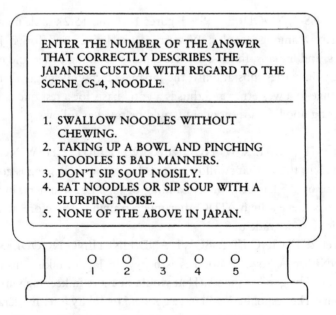

Figure 15.2   Answers for noodles scene.

► Answer 3: DON'T SIP SOUP NOISILY. Eating without noise is deemed most polite by westerners, but Japanese don't value it. The respondent needs a lot of training and is graded 3.

► Answer 4: EAT NOODLES OR SIP SOUP WITH A SLURPING NOISE. Japanese like to make noise while drinking tea, coffee, or soup. In the tea ceremony, specifically, the last sip should be accompanied with a sipping noise. Therefore, the respondent is a person who understands Japanese culture and is graded 1.

► Answer 5: NONE OF THE ABOVE IN JAPAN. The respondent has not found an appropriate answer among the alternatives. Thus he or she should have a very low ATDC and is graded 4.

Example 2: CS-11, PORCH.   (See Figures 15.3 and 15.4.) We invited a foreign couple to visit us. They came to our house and entered through the porch. I told them "Please come in." The lady then stepped inside without taking her shoes off. Usually, in Japan people go barefoot in the house. One of the greatest sources of embarrassment is that a foreigner is used to going into a house with shoes on. According to the program, a respondent will be rated as follows:

Figure 15.3   Scene: at the porch.

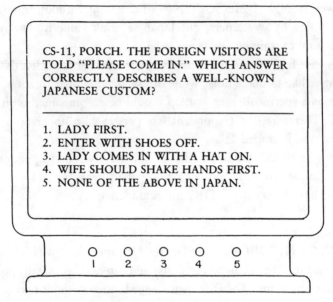

CS-11, PORCH. THE FOREIGN VISITORS ARE TOLD "PLEASE COME IN." WHICH ANSWER CORRECTLY DESCRIBES A WELL-KNOWN JAPANESE CUSTOM?

1. LADY FIRST.
2. ENTER WITH SHOES OFF.
3. LADY COMES IN WITH A HAT ON.
4. WIFE SHOULD SHAKE HANDS FIRST.
5. NONE OF THE ABOVE IN JAPAN.

O   O   O   O   O
1   2   3   4   5

Figure 15.4   Answers for porch scene.

▶ Answer 1: LADY FIRST. The respondent receives an ATDC grade of 2.

▶ Answer 2: ENTER WITH SHOES OFF. This answer accurately reflects Japanese custom and will be given a grade of 1.

▶ Answer 3: LADY COMES IN WITH HER HAT ON. The respondent is graded 3.

▶ Answer 4: WIFE SHOULD SHAKE HANDS FIRST. The respondent is graded 3.

▶ Answer 5: NONE OF THE ABOVE IN JAPAN. This answer is given the lowest grade, 4.

Example 3: CS-19, SLIPPERS.   (See Figures 15.5 and 15.6.) In a Japanese house, people usually walk barefoot or wear either socks or *tabi,* Japanese-style short socks. Slippers are offered to guests to keep their socks clean. This causes another problem, because foreigners are not used to wearing such slippers in their home country. They usually wear shoes at home or when visiting. Furthermore, for the lavatory, Japanese provide a separate pair of slippers to be worn exclusively there. Because the Japanese regard the toilet as a dirty

Figure 15.5   Scene: SLIPPERS.

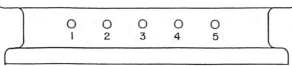

Figure 15.6   Answers for SLIPPERS scene.

place, these special slippers are never used outside the toilet room. Therefore, everybody is requested to change slippers upon entering the lavatory.

In a room where *tatami* (Japanese straw mats) are laid, the Japanese do not usually use slippers, because they don't like to see any dust that may be stuck to the underside of the slippers drop on the mats. It is troublesome to have to sweep the *tatami* later. Therefore, everybody is asked to remove his or her slippers when entering a room with *tatami*. This is a characteristic Japanese custom.

Using the grading system, the ATDC scores for Figures 15.5 and 15.6 are as follow:

▶ Answer 1: CHANGE OR TAKE OFF SLIPPERS WHEN ENTERING LAVATORY OR TATAMI ROOM. This is the Japanese custom. It receives an ATDC grade of 1.

▶ Answer 2: TO KEEP SOCKS CLEAN, PUT ON SLIPPERS. This partially describes the Japanese custom and is graded 2.

▶ Answer 3: PUTTING ON AND TAKING OFF SLIPPERS SUBJECT TO ROOM FLOOR IS TROUBLESOME. The respondent shows great uncertainty and is graded 3.

▶ Answer 4: ENTER THE HOUSE WITH SHOES ON. The respondent is again graded 3.

▶ Answer 5: NONE OF THE ABOVE IN JAPAN. As before, respondent receives a grade of 4.

We prepared computer programs for this test and exhibited the system at a symposium in 1988.

Computer Analysis

We give a test comprising 30 scenes before a training program and another test after the program. The comparison of ATDC scores tells us individuals' rates of progress in cultural adaptation, while the average score difference tells us the effectiveness of the training. A more advanced course in cultural simulation can be taken afterward if desired.

By testing and improving their ATDC, foreign-born engineers can have an enjoyable stay and a productive working life in Japan.

The Benefits of Professional Societies

16

The Benefits of American Professional Societies, their Activities in Japan, and the International Congress of Mechanical Engineering Societies (ICOMES)

Hiroshi Honda

INTRODUCTION

European professional societies have enjoyed leading roles in academia. Their influence was particularly strong in the periods just preceding World Wars I and II. The wars had a significant detrimental effect on the European professional societies. Since the end of World War II, many American professional societies took their place in the leading roles in the international arena, and the American societies' activities such as hosting international conferences and technical presentations spread around the globe. In the half century since the war, the increasing economic interdependence among the countries of the world has increased the frequency of such international events and conferences. As Japan emerged as the second largest economic power in the world beginning in the 1980s, many international conferences of American societies came to be held in Japan, leading to enhanced interaction and friendship between the two countries and with other countries participating in those activities. The booming economy in the Asia-Pacific region beginning the 1990s also promoted such international activities, especially those of the Asia-Pacific Economic Cooperation (APEC), ranging from summit meetings to technical seminars and workshops in the Asia-Pacific region. These have led to the formation of international congresses and societies in various disciplines, such as the mechanical engineering societies introduced in the next part of this chapter.

Some of the leading professional societies, such as the American Society of Mechanical Engineers (ASME) and the American Society of Civil Engineers (ASCE), established Japan chapters in the 1980s which were later upgraded to sections. These chapters provided benefits for Japanese members not available through the societies' Japanese counterparts, such as the seminars on boiler, pressure vessel, and piping codes and standards and on project management methods in overseas countries.

Many American universities established Japan clubs and/or Japan chapters of their alumni associations in the 1980s, leading to increased opportunities for American and Japanese people in academic, business, and governmental sectors to meet and discuss various matters of mutual interest. Some eminent American universities established industry-academia programs, such as the Mitsubishi-Harvard and the Mitsui-MIT programs in Japan, whose agendas are oriented toward the interests of the companies within these groups. Some universities established Japanese extensions, such as the Stanford Center in Kyoto and the Temple University campus in a Minato-Ku, Tokyo. Universities such as Penn State have established relationships in the materials science discipline with leading Japanese companies in the steel, glass and ceramics, and electronics materials sectors. Such relationships can be seen between many other American universities and Japanese industries; however, the economic slump that began in the 1990s has limited available funds for maintaining these relationships in Japan.

This chapter will describe the benefits of being affiliated with American and international professional societies, citing several examples from recent decades.

AMERICAN PROFESSIONAL SOCIETIES

Among the many American professional societies that have played leading roles in their own disciplines, the American Society of Mechanical Engineers (ASME) International, which serves as secretariat for the International Congress of Mechanical Engineering Societies (ICOMES), is a case in point.

ICOMES and ASME International

The International Congress of Mechanical Engineering Societies (ICOMES) was first convened in May 1987 to provide a forum on issues important to technical societies, as well as to encourage international cooperation among participating societies. The 57 mechanical engineering societies of 56 nations around the globe have participated in ICOMES as of September 1, 1999. The ICOMES newsletter is also released four times a year, and the online version of the newsletter, *ICOMES Online,* is available at *http://www.asme.org/international/icomes* on a regularly updated basis.

Among these societies, ASME International is the largest and probably the most active society in the world in the mechanical engineering discipline. ASME International enjoys by far the largest membership, about 130,000, followed by the Institution of Mechanical Engineers (IMechE) of the United Kingdom and the Japan Society of Mechanical Engineers (JSME). The range of activities of ASME's 38 technical divisions is extensive; these are compared with the activities of JSME's 20 technical divisions in Table 16.1. The schedules for a variety of international and ASME committee meetings, symposia, and conferences are available at *http://www.asme.org/NS/Search/ASMECalendar.* The international

Table 16.1 Divisions of the American Society of Mechanical Engineers (ASME) and the Japan Society of Mechanical Engineers (JSME)

ASME	JSME	日本機械学会 (in Japanese)
Advanced Energy Systems		
Aerospace	Space Engineering	宇宙工学
Applied Mechanics	Materials and Mechanics	材料力学
Bioengineering	Bioengineering	バイオエンジニアリング
Computers and Information in Engineering	Computational Mechanics	計算力学
Design Engineering	Design and Systems	設計工学・システム
Dynamic Systems and Control	Dynamics and Control	機械力学・計測制御
Electrical and Electronic Packaging	Robotics and Mechatronics	ロボティクス・メカトロニクス
Environmental Engineering	Environmental Engineering	環境工学
Fluids Engineering	Fluids Engineering	流体工学
Fluid Power Systems and Technology		

(continues)

Table 16.1   *(continued)*

ASME	JSME	日本機械学会 (in Japanese)
Fuels and Combustion Technologies		
Heat Transfer	Thermal Engineering	熱工学
Heat Transfer/K16 (H.T. in Electronic Equipment)		
Information Storage and Processing Systems	Information, Intelligence and Precision Equipment	情報・精密機械
Internal Combustion Engine	Engine Systems	エンジンシステム
International Gas Turbine Institute		
Management		
Manufacturing Engineering	Factory Automation	ファクトリーオートメーション
Materials	Manufacturing and Machine Tools	加工学・加工機器
	Materials and Processing	機械材料・材料加工
Materials Handling Engineering		
Noise Control and Acoustics		
Non-Destructive Evaluation Engineering		
Nuclear Engineering		
Ocean Engineering	The Society of Naval Architects, Japan (a separate, independent Japanese society)	日本造船学会
Offshore Mechanics and Arctic Engineering		
Petroleum		
Plant Engineering and Maintenance		
Power	Power and Energy Systems	動力
Pressure Vessels and Piping		
Process Industries	Industrial and Chemical Machines	産業・化学機械
Rail Transportation	Transportation and Logistics	交通・物流
Safety Engineering and Risk Analysis		
Solar Energy		
Solid Waste Processing		
Technology and Society	Technology and Society	技術と社会
Textile Engineering		
Tribology, Gear Research Institute	Machine Design and Tribology	機素潤滑設計

activities, including those in Japan, elsewhere in Asia, and in other countries, are also announced in the *ASME Worldwide Newsletter*.

Many international engineers, scientists, and other professionals have chances to get acquainted with Japanese professionals through technical and other presentations, and some of these interpersonal contacts could sooner or later lead to an opportunity to work in Japan.

Activities of American Professional Societies in Japan

ASME International has enjoyed an excellent friendship with JSME since the two organizations concluded an agreement of cooperation in 1979. Among all international societies in Japan, American societies would be among the most active, owing much to an increasing economic, cultural, scientific, and technological interdependence between Japan and the United States. English-speaking engineers and scientists with limited Japanese language skills will probably benefit most by joining American societies such as the Japan Section of the American Society of Mechanical Engineers (whose chairman for the 1998–2000 term is Professor Junjiro Iwamoto of Tokyo Denki University). ASME Japan was established in 1986 and took its place alongside other American scientific, technological, and professional societies in Japan that have acted as catalyst and lubricant in promoting joint activities between parent societies in the United States and their Japanese counterparts, as well as between Japanese companies and universities. In January 2000, ASME President Robert Nickell and Executive Director David Belden visited Japan, and ASME Japan hosted a dinner reception in their honor. Since the activities of this society are introduced elsewhere in this book, details are omitted here. Other societies (such as the Japan Section of the American Society of Civil Engineers) were also established in the 1980s and have continuing friendships with corresponding societies in Japan (such as the Japan Society of Civil Engineers).

Activities of American Universities in Japan

Alumni associations of many American universities have clubs or chapters in Japan. The Penn State Alumni Association, which has the largest membership of any alumni organization in the United States, and other alumni organizations, such as those of Harvard, Stanford, the University of Illinois, and Dartmouth

College, have active chapters or clubs in Japan. These universities have produced numerous Japanese business and academic leaders, bureaucrats and diplomats, and Olympic medalists in wrestling, skiing, and other sports since the Meiji era, when Japanese began to go mainly to the United States and Europe as students, research associates, visiting scholars, and visiting professors.

These alumni associations host annual and special meetings in honor of eminent scholars from their alma maters, and the connections obtained through this kind of activity can directly or indirectly benefit foreign-born professionals working in Japan.

SUMMARY REMARKS

American professional societies carry on many activities that provide benefits to foreign-born professionals. But one should not expect specific results from these activities, since rewards often come when people least expect them. One should feel content just to relax and enjoy the activities while living away from the home country. Deepening your acquaintance with friends, senior advisers, and mentors in your own and/or different disciplines can expand your capacity and your horizons as both a human being and a professional.

17

The Benefits of Other International Societies in Japan

Hiroshi Honda

Japan first initiated diplomatic intercourse with China and Korea about 2000 years ago, and with Portugal in 1543; in the course of their first contacts, the Portuguese brought guns to Japan. Thereafter, other Europeans, such as the British and Dutch, brought their own culture and goods to Japan, even though Japan's contacts with these countries were limited by its policy of national isolation during the Edo period from 1633 to 1867. In 1867, Japan adopted its policy of opening to the world, and intercourse with foreign countries has since accelerated, especially after the end of World War II. Today, there are many oriental and European organizations and societies in Japan that would benefit foreign-born professionals working there.

Organizations such as the China-Japan, the Korea-Japan, the Brazil-Japan, and the Peru-Japan friendship associations have facilitated bilateral relationships as Japan has increasingly accepted employees of Japanese and other origin from these nations, among others. Japan has also enjoyed multinational as well as bilateral friendships with western peoples and nations through the European Union, the British Council, the Center National de la Recherche Scientifique (CNRS)–Japon (Japan Chapter of the National Center for Scientific Research, France), and the Deutsches Institute für Japanstudien (German Institute for Japanese Studies), among others. The alumni associations of universities such as the Australian National University and various European universities have facilitated international activities through international meetings and scholarship programs.

The European Union, for example, promotes international communication between Europe and Japan through exchange programs for scholars, scientists, and technologists, by hosting and participating in joint events with the Japanese, and through public and government-affiliated organizations. For example, the Japan Space Utilization Promotion Center (JSUP) regularly invites scientists and technologists from the European Space Agency, the French Space Agency (CNES), the German Space Agency, the Russian Space Agency, and the National Aeronautics and Space Administration (NASA), among others, to its annual International Symposium on Space Utilization (IN SPACE symposium).

The British Council (BC) hosts activities and events in over 100 countries and has a Japan office in Shinjuku, Tokyo (the nearest JR access is from the Iidabashi station). The BC is open to anyone for an annual membership fee of 4000 yen, as of 1999. Information services such as Internet access, video and audio cassette tapes, and the library are available free of charge; and various events on the arts and sciences hosted by the BC, such as conferences, seminars, lectures, and film nights, as well as a counseling service, are open to members at a discount rate. The CNRS–Japon and the Deutsches Institut für Japanstudien also have offices at, respectively, the French-Japanese Hall in Ebisu, Shibuya-ku, Tokyo, and Kudan Minami, Chiyoda-ku, Tokyo. Japanese and international professionals are often invited to participate in their events, such as conferences and lectures with guest scholars and other authorities from France and Germany.

The Kaisha Society was formed mainly by Americans and Europeans who work at Japanese companies. It hosts dinner receptions with distinguished speakers, such as ambassadors to Japan from the western nations; conducts surveys on working conditions and salaries among foreign-born professionals in Japan; and hosts many other events. The society should be beneficial to English-speaking professionals in Japan.

The aforementioned societies offer excellent opportunities for foreign-born professionals to enjoy rest and recreation amid an international atmosphere at a modest cost, or often free of charge. These societies offer a mixture of things Japanese, oriental, and western, and thus are distinct from their counterparts back home.

18

The Benefits of Japanese Professional Societies

Hiroshi Honda

INTRODUCTION

Foreign-born professionals are encouraged to participate in the events and meetings hosted by Japanese professional societies, where they can fully enjoy things Japanese and oriental. They can participate in the goings-on using the Japanese language, which they often need not speak fluently, or in English or other foreign languages or a mixture. Japanese people appear to be shy at first, but are willing to communicate to the best of their ability with their international colleagues, once they recognize that they have a common interest on which they can exchange views and opinions. They might even enjoy communicating in a mixture of broken languages—often sufficient for effective communication on topics of everyday interest. This willingness, of course, depends on the individual personality, just as it does among people of any nationality.

THE JAPAN SOCIETY OF MECHANICAL ENGINEERS (JSME) AND OTHER ACADEMIC SOCIETIES

JSME is considered one of the most prestigious academic societies in Japan, its membership of 45,000 being about the largest among them. JSME was established in 1897, when the second university-level mechanical engineering department in the nation was established at Kyoto University, following its counterpart at the University of Tokyo. JSME is a little more bureaucratic and academia-oriented than ASME International in general; however, JSME emphasizes its policy on internationalizing the society, participates in ICOMES activities, and has concluded agreements of cooperation with many other engineering societies around the world.

JSME publishes the monthly *Journal of the Japan Society of Mechanical Engineers* in Japanese, covering general topics on mechanical engineering. JSME also publishes the monthly English-language *JSME International Journal,* which contains technical papers, and distributes it throughout the world; distribution in North and South America is handled by ASME International through a cooperation agreement. An English version of *JSME News* is also published, and *Transactions of the JSME* contains English abstracts, and English table and figure captions to accompany its papers in Japanese. It also holds joint conferences with ASME, the Korean Society of Mechanical Engineers, and many other mechanical engineering societies, in Japan and in many cities outside of Japan. Details of its activities are available at its Web site, *http://www.jsme.or.jp/.*

There are also numerous other academic societies in Japan in a variety of disciplines, such as the Chemical Society of Japan, the Institute of Electronics, Information and Communication Engineers, the Japan Society of Civil Engineers, the Architectural Institute of Japan, the Japan Society of Microgravity Application, and the Japanese Society for Biological Science. These societies also sponsor and cosponsor international conferences and events in English.

JAPAN SOCIETY FOR THE PROMOTION OF SCIENCE (JSPS) ____

JSPS is a Monbusho-affiliated organization, as noted in Chapter 4, headquartered in Koji-machi, Chiyoda-ku, Tokyo, with Hiroyuki Yoshikawa, the former president of the University of Tokyo, as its chairman and Ken Kikuchi as president. Information on JSPS can be accessed at its Web site, *http://www.jsps.go.jp.* To assist and promote its extensive international activities around the world, it also has liaison offices in Washington, D.C., in the United States; Bonn, Germany; London, United Kingdom; Bangkok, Thailand; Cairo, Egypt; Nairobi, Kenya; and São Paulo, Brazil.

JSPS carries out and administers the following international programs, in addition to their domestic counterparts:

1. Invitation fellowship programs for research in Japan, and postdoctoral fellowship programs for foreign researchers and for research abroad.
2. International scientific meetings
3. Bilateral programs with 38 nations, one region, and 61 organizations around the world as of F.Y. 1997
4. Cooperative programs with Asian countries
5. An international prize for biology
6. Activities carried out by seven liaison offices around the world
7. Programs to assist international fellowship awardees at JSPS Fellow Plaza (address, Jochi Kioizaka, 6-2-3, Kioicho, Chiyoda-ku, Tokyo 102-0094; telephone, 81 + (0)3-3263-1721; fax, 81 + (0)3-3263-1854)

Foreign-born scholars residing in Japan are advised to contact JSPS offices for any inquiries concerning professional work, everyday life, family considerations, and cultural and language issues and lessons.

SCIENCE- AND TECHNOLOGY-RELATED CENTERS, FOUNDATIONS, INSTITUTES, AND INSTITUTIONS IN JAPAN ___

As of October 1, 1998, 115 juridical persons* had joined the United Science and Technology Organizations (Kagaku-Gijutu Dantai Rengo), Japan, and 15 others who were under the administration of the Science and Technology Agency had not. The United Science and Technology Organizations are headed by Eishiro Saito, chairman of the Japan Science Foundation (JSF), with its secretariat at the Japan Foundation of Public Communication on Science and Technology, or JFPCST (Minoru Oda, chairman, and Masaaki Kuramoto, president).

Among the organizations, there are 39 science- and technology-related centers, foundations, institutes, and institutions that carry on international activi-

---

* Legal term meaning organizations legally established and registered, such as foundations, companies, ect.

ties, and 26 that offer national and international awards and citations. Their activities are a potential benefit to foreign-born engineers and scientists. Inquiries can be made to the JFPCST (address, 1-8-6, Akasaka, Minato-ku, Tokyo 107-0052; telephone, 81 + (0)3 3586-0681; fax, 81 + (0)3 3586-0686; and e-mail, *pcst@venus.dti.ne.jp*).

INDUSTRIAL SCIENCE- AND TECHNOLOGY-RELATED CENTERS AND FOUNDATIONS IN JAPAN

There are 17 centers, associations, institutes, agencies, and partnerships in the private sector, as shown in Table 18.1, that implement research and development in industrial science and technology and are administered by the Agency of Industrial Science and Technology of MITI and the New Energy and Industrial Technology Development Organization (NEDO) in the public sector. Those private-sector organizations, shown in Table 18.1, besides conducting research and development, host seminars and/or symposia jointly with related MITI-affiliated national research institutes and laboratories from among those listed in Figure 1.5. Industrial technology fellowship awardees are recruited by NEDO (1-1, 3-chome, Higashi Ikebukuro, Toshima-ku, Tokyo 170; telephone, 81 + (0)3 3987-9354; fax, 81 + (0)3 3987-1536), which in turn sends them to the organization(s) in charge of specific research projects.

THE JAPAN EXTERNAL TRADE ORGANIZATION (JETRO); THE INSTITUTE OF ENERGY ECONOMICS, JAPAN (IEEJ); AND NUMEROUS OTHER ORGANIZATIONS ADMINISTERED BY MITI

MITI administered over 900 public and private-sector organizations as of March 1999, including JETRO (public sector) and IEEJ (private sector), that can benefit engineers, scientists, and businesspeople through events such as international exhibitions, conferences, and seminars held in Japan and overseas countries.

JETRO hosts the technology tie-up promotion program to facilitate technical licensing and other agreements between overseas and Japanese companies.

Table 18.1 Organizations Implementing R&D in Industrial Science and Technology

Organization Name	Address	Telephone (Upper) and Fax (Lower) Numbers	Project under supervision
International Superconductivity Technology Center	5-34-3, Shinbashi, Minato-ku, Tokyo 105	(03) 3431-4002 (03) 3431-4044	Superconducting materials and devices
Research and Development Association for Future Electron Devices	8-10-24, Akasaka Minato-ku, Tokyo 107	(03) 3423-1621 (03) 3431-1680	Superconducting materials and devices and quantum functional devices
R&D Institute of Metals and Composites for Future Industries	3-25-2, Toranomon, Minato-ku, Tokyo 105	(03) 3459-6900 (03) 3459-6911	High-performance materials for severe environments
Japan High Polymer Center	2-5-3, Kuramae, Taito-ku, Tokyo 111	(03) 3851-8860 (03) 3864-4627	Nonlinear photonics materials, silicon-based polymers
Advanced Chemical Processing Technology Research Association	10-6, Hisamatsu-cho, Nihonbashi, Chuo-ku, Tokyo 103	(03) 3661-6561 (03) 3661-6870	Advanced chemical processing technology
Fine Ceramics Research Association	3-7-10, Toranomon Minato-ku, Tokyo 105	(03) 3437-3651 (03) 3437-3650	Synergy ceramics
Marine Biotechnology Institute	2-35-10, Hongo, Bunkyo-ku, Tokyo 113	(03) 5684-6211 (03) 5684-6200	Fine chemicals from marine organisms and tropical bioresources
Research Association for Biotechnology	2-3-9, Nishi-Shinbashi, Minato-ku, Tokyo 105	(03) 3595-0371 (03) 3595-0374	Molecular assemblies for a functional protein system, etc.
Information Technology Promotion Agency, Japan	3-1-38 Shibakoen, Minato-ku, Tokyo 105	(03) 3433-2350 (03) 3437-5386	New models for software architecture

(continues)

Table 18.1 (*continued*)

Organization Name	Address	Telephone (Upper) and Fax (Lower) Numbers	Project under supervision
Angstrom Technology Partnership	2-5-12, Higashi kanda Chiyoda-ku, Tokyo 101	(03) 3423-1621 (03) 3423-1680	Ultimate manipulation of atoms and molecules
Femtosecond Technology Research Association	5-5, Tokyo-dai, Tsuku-ba, Ibaraki 300-26	(0298) 47-5181 (0298) 47-4417	Femtosecond technology
Engineering Research Association for Super/Hyper Transport Propulsion System	2-6, Kohinata 4-Chome, Bunkyo-ku, Tokyo 112	(03) 5684-5180 (03) 5684-7540	Super/hypersonic transport propulsion systems
Micromachine Center	2-2, Kanda-Tsukasa-cho, Chiyoda-ku, Tokyo 101	(03) 5294-7131 (03) 5294-7137	Micromachine technology
Technology Research Association of Ocean Mineral Resources Mining System	5-11-13, Ginza, Chuo-ku, Tokyo 104	(03) 3542-6091 (03) 3541-6840	Basic technology, manganese nodule mining technology
Engineering Advancement Association	1-4-6 Nishi-Shinbashi Minato-ku, Tokyo 105	(03) 3502-3671 (03) 3502-3265	Underground space development technology
Research Institute of Human Engineering for Quality Life	1-2-5, Dojima, Kita-ku, Osaka 530	(06) 346-0234 (06) 346-0456	Human sensory measurement application technology
Technology Research Association of Medical and Welfare Apparatus	3-5-8, Shiba Koen, Minato-ku, Tokyo 105	(03) 3459-9584 (03) 3459-6887	Medical and welfare

*Note: The organizations listed here are not national organizations and thus are not listed in Appendix 2.

JETRO is headquartered at 2-2-5, Toranomon, Minato-ku, Tokyo 105-8466 (telephone, 81 + (0)3 3582-4631 and fax, 81 + (0)3 3582 7508 for its Industrial Cooperation Division, and Web site *http:/www.jetro.go.jp/*), in the vicinity of the American embassy in Tokyo. JETRO headquarters has a library open to visitors, with publications on trade, economy, and industrial issues. JETRO also hosts events through its more than 80 offices and centers over the world, and accommodates booths at international exhibitions such as the annual World PC Expo, the largest of its kind in Asia, to give representatives of overseas companies an opportunity to introduce their products to Japanese people.

IEEJ is the most active organization in Japan in the field of energy economics; it comes under the administration of the Agency of Natural Resources and Energy of MITI and is the parent organization of the Energy Data and Modeling Center, the Petroleum Information Center, and the Asia-Pacific Energy Research Centre (APERC). IEEJ has assisted MITI in forming Japan's energy policy for more than 30 years, including the period of the two oil crises in the 1970s, and has a continuing relationship with the International Energy Agency of OECD and the U.S. Department of Energy, among others. IEEJ annually hosts the Symposium on Pacific Energy Cooperation (SPEC), the APEC Coal Flow Seminar, and the Japanese Committee for Pacific Coal Flow (JAPAC) International Meeting, among others, to offer opportunities for dialogue between energy producers and consumers around the world. IEEJ and APERC accept visiting researchers from the government and from government-related energy organizations in the Asia-Pacific region and other regions. Foreign professionals can obtain APEC energy-related publications from APERC upon request, and their participation in the aforementioned symposia is welcomed.

There are many other MITI-related organizations that can benefit foreign professionals. For details concerning the activities of these organizations, visit MITI's Web site at *http://www.miti.go.jp/*.

LOCAL INTERNATIONAL (FRIENDSHIP) ASSOCIATIONS _____

Local volunteer-based international associations, a majority of them sponsored by the municipal and prefectural governments, often exist to provide (often free of charge) Japanese-language lessons, cultural events and seminars, interpreting services, and various other activities for foreign-born residents in Japan. These associations regularly hold exchange programs with counterparts (if they exist) in sister cities in countries such as the United States. For example, Narashino International Association (NIA) and Urayasu International Friendship Association (UIFA), with Tokyo Disneyland in its city, in Chiba prefecture holds a sister relationship with the counterparts of Tuscaloosa, Alabama and Orlando, Florida respectively. Their membership fee is set at a very modest level of around 1000 yen to 2000 yen a year. Guidebooks for living in local cities and prefectures are often provided in English, Spanish, and Chinese. Foreign-born residents should contact the local government offices of their place of residence to find out the names of contacts (who often themselves have international experience, having resided outside of Japan) and their addresses and phone numbers and the activities they take part in.

SUMMARY REMARKS

There are many other public- and private-sector organizations administered by other governmental organizations, such as the Ministry of Transport, the Ministry of Posts and Telecommunications, the Ministry of Construction, the Ministry of Agriculture, Forestry and Fisheries, the Ministry of Finance, the Ministry of Justice, the Environment Agency, the Ministry of Labor, and the Economic Planning Agency. These organizations offer similar events, within the scope of their own respective activities, and should benefit foreign professionals who work in related disciplines.

19

Toward Mutual Recognition of Engineering and Technology Education and Qualification between the United States, Japan, and Rest of the World: The Role and Efforts of ABET and JABEE

Winfred M. Phillips, George D. Peterson, Kathryn B. Aberle, and Hiroshi Honda

INTRODUCTION

As industrial operations have become more international and national economies have become more closely bound to a global economy, engineers and scientists have been enjoying increasing opportunities to collaborate with their counterparts in other nations. In order to cope with this trend, the effort to establish global qualification standards in the engineering profession and accreditation standards for engineering and technology education has accelerated.

The Accreditation Board for Engineering and Technology (ABET) in the United States was a leader in this effort and in 1989 signed the Six Nation Accord (which later became the Washington Accord), under which Australia, Canada, Ireland, New Zealand, the United Kingdom, and the United States formally recognized that their basic engineering education meets similar standards. The Engineering Council of South Africa and the Hong Kong Institute of Engineers became signatories to the accord in 1993 and 1995, respectively. In 1999, the preparation committee for the establishment of the Japan Accreditation Board for Engineering Education (JABEE) declared that it was aiming to become a signatory to the Washington Accord in the early 2000s. Nations such as Argentina and France signed memoranda of understanding (MOUs) concerning the development of accreditation procedures similar to the ones ABET uses, such as peer review.

This chapter focuses on the role and efforts of ABET and JABEE in promoting mutual recognition of engineering and technology education and qualification of engineers in the global community. The chapter takes particular note of the ways their activities will benefit engineers and scientists already working in or planning to work in Japan.

ACCREDITATION BOARD FOR ENGINEERING AND TECHNOLOGY (ABET), U.S.A.[1]

ABET, founded in 1932, and its predecessor, the Engineers' Council for Professional Development (ECPD), have been responsible for the assurance of quality in engineering education in the United States. ABET is a federation of 27 professional engineering and technical societies that have joined together to promote and enhance engineering, technology, and applied science education. It boosted the quality and credibility of U.S. engineering programs but raised some concerns that its rigorous criteria also resulted in too much standardization of engineering programs. In response, ABET has launched a revolutionary makeover of its accreditation procedure aimed at facilitating innovation and creativity. Engineering Criteria 2000 (EC2000) changes the emphasis of the accreditation evaluation from what is taught to what is learned. ABET also expanded its mission to include accreditation of applied science programs and will have a new commission to accredit computing science programs in 2001, following an agreement with the Computing Sciences Accreditation Board.

An integral part of current ABET policy is to emphasize international engineering educational quality assurance through the development of accreditation systems. ECPD worked closely with the Canadian engineering communities from 1940 to the 1970s to assist them in developing their own

---

[1] Winfred M. Phillips, George. D. Peterson, and Kathryn B. Aberle, *Quality Assurance for Engineering Education in a Changing World,* publication in progress.

accreditation systems, and ABET, with Mexican engineering communities in the 1990s to help them develop criteria for evaluation of engineering programs. In addition to the nations previously listed as signatories to the Washington Accord, ABET signed MOUs with the Regional Office for Science and Technology for Latin America and the Caribbean of the United Nations Educational, Scientific and Cultural Organization (UNESCO) in 1995.[2] ABET has also been in contact with representatives of Ukraine, Russia, and Germany, among others, to facilitate international cooperation.

ABET formalized the procedures and policies used to evaluate programs at institutions outside the United States in 1991 to deem that they are "substantially equivalent" to those accredited in the United States, and has evaluated and recognized over 70 programs at 14 institutions in 10 countries.[3] The ABET International Activities Committee is currently examining ways in which extension of EC2000 evaluations beyond U.S. borders can be accomplished, and this will require the countries to have an educational process comparable to that in the United States. In the fall of 1997, ABET established a service to evaluate the educational credentials of engineers who attended institutions outside the United States, and Engineering Credentials Evaluation International (ECEI) evaluates credentials using the ABET engineering criteria in effect at the time the applicant graduated. ABET intends to continue partnering with organizations around the world to ensure that engineers receive the education they deserve and that the profession and the public can rely on that level of educational quality.

ESTABLISHMENT OF THE JAPAN ACCREDITATION BOARD FOR ENGINEERING EDUCATION (JABEE)

Japanese college engineering education was by and large dependent on individual universities' policies until the early 1990s. For example, engineering education at universities in western Japan was generally influenced by its German counterpart, while that at universities in eastern Japan was more influ-

---

[2] Memorandum of Understanding between ABET and UNESCO, November 15, 1994.
[3] 1997 ABET International Yearbook, Accreditation Board for Engineering and Technology, Baltimore, Md., 1998.

enced by British and American counterparts. Doctor's degrees in engineering, called kogaku hakushi, conferred by the universities, particularly in western Japan, were equivalent to the German Dr. Ing. degree, with a very rigorous dissertation requirement. The Japanese qualification system for professional engineers was also based on a requirement for significant experience in the engineering professions, and Japanese engineers usually acquired a license as Registered Consulting Engineering (*gijutsushi*) only in their late thirties or in their forties. In contrast, professional engineer's licenses in the United States, the United Kingdom, France, and other nations influenced by them are usually acquired at younger ages.

With the globalization of the engineering profession and a move toward mutual recognition of academic and professional qualifications in the global community, the Japanese universities adopted the Ph.D. (hakushi) system in the 1990s. Organizations such as the Engineering Advancement Association of Japan (EAAJ) investigated the qualification systems of the engineering professions around the world in the 1990s and discussed policies and measures to comply with global standards. In 1998, a system for accreditation of engineering and technology education compatible with global standards began to be seriously considered on a national scale.[4] A preparation committee for the establishment of JABEE was officially set up on April 30, 1999, with the support of the organizations administered by Monbusho (Ministry of Education, Sports, Science, and Culture), and JABEE is expected to be officially established in the near future.

The preparation committee for the establishment of JABEE clearly states that it will make every effort to sign the Washington Accord by 2004. The committee began studies on establishing JABEE standards for engineering ethics, engineering education, and the role of science in engineering education, among other issues, and determined that JABEE shall have the following missions:[5]

1. Accredit engineering education programs
2. Establish accreditation standards

---

[4] Hiroyuki Yoshikawa, "Issues on the Accreditation System for Engineering Education and the Qualification of Engineers," speech by the President of the Science Council of Japan, December 17, 1998 (given in Japanese).

[5] Hideo Ohashi, Establishment of Japan Accreditation Board for Engineering Education (JABEE) (in Japanese), downloaded from *(http://www.kanazawa-it.ac.jp/JABEE/)* on December 25, 1999.

3. Publicize its accreditation activities in Japan
4. Conduct studies and surveys concerning accreditation
5. Nurture and educate specialists in accreditation
6. Publicize its activities to overseas nations
7. Form liaisons with Washington Accord–related organizations such as ABET, and coordinate on accreditation issues

Japan, Australia, the Philippines, and Indonesia are the core members of the project for recognition of qualification of engineers within the APEC region, which was initiated within the APEC Human Resources Working Group; and Thailand, Korea, New Zealand, and the United States are also members of the project. JABEE is therefore expected to become a key organization in establishing the recognition system, introducing a qualification such as APEC Engineer, collaborating with the other overseas counterpart organizations.

Japan also has a keen interest in the service trade and related issues being discussed and negotiated within the World Trade Organization (WTO), which would have a substantial linkage with recognition of qualifications in various professions, including engineers. There will be increasing opportunities for Japan and counterpart organizations in the United States to collaborate on these issues.

TOWARD MUTUAL RECOGNITION ON REGIONAL AND GLOBAL SCALES

Engineering mobility has become an especially important issue in regional communities such as those of the North American Free Trade Agreement (NAFTA), APEC, and the European Union (EU). The move toward mutual recognition for engineering and technology education and qualification is also steadily on the rise on a global scale, as can be seen in activities such as those of the WTO and UNESCO. There is no doubt that ABET will play a key role on a global scale as a representative of the United States, and JABEE is expected to play a similar role in Japan and the APEC community toward mutual recognition. It is worthwhile for engineers working abroad to watch these trends, and specifically for those working or planning to work in Japan to participate in and contribute to JABEE activities.

Employment
Case Studies

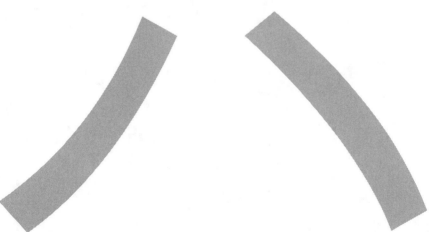

20

Perspectives of Industry

P. E. D. Morgan, Charles R. Heidengren, and Dimitrios C. Xyloyiannis

IMPRESSIONS FROM A ROCKWELL-HITACHI EXCHANGE PROGRAM (1989) AND LATER WORK EXPERIENCES (TO 1999)

Peter E. D. Morgan

The Hitachi Visit and Rockwell-Hitachi Exchange Program

I had visited Japan four times, for periods of about two weeks each, before I went in 1989 on an exchange program, to work for six months at the Super-conductivity Center of the Hitachi Corporation in Hitachi-shi (city), Ibaraki-ken (prefecture), located about 150 kilometers northeast of Tokyo on the Pacific coastline. My earlier visits had primarily been for scientific conferences, but each time I had spent about one week touring Japan and visiting colleagues, some of fairly long standing. I had taken a very short course in Japanese (spoken only) at Rockwell, mainly out of curiosity about how the structure of the language differed from that of the Indo-European group. I found Japan to be extremely intriguing in having attained success with a different mix of assumptions from those implicit in the western model and was greatly interested in the different weight given to ancient and modern ideas and traditions. To realize that religious concepts and customs akin to those of the Greek and Roman civilizations—the western fountainheads—are alive and well today,[1] coexisting with influences from India, Korea, and China, ... that's a long story.

---

[1] Space prevents detail, but modern manifestations are, for example, the Shinto blessing and inauguration ceremonies for the installation of important new pieces of scientific equipment and for the sealing of significant business contracts.

At school in my native Great Britain, I had studied French and Latin and so was not nervous about attempting to learn another language; in fact, I regarded it as a challenge, anticipating the language, as is usual, to embody something of what makes its associated culture distinct. In any event, what I had learned from the trips, and the pathetically little language I already knew, encouraged me toward more effort; reading literature on Japan, I was particularly impressed by the writings of Lafcadio Hearn (alias Yakumo Koizumi), a rather curious Britisher (of Irish-Greek extraction) who arrived in Japan in 1890, never again to leave, and who insightfully documented Japanese customs, mores, folktales, and ghost stories (learned from his wife), among other things. He wrote in English, I believe, but now, translated back into Japanese, his work is part of the lore of Japan. I felt an accord with his observations that reflected, I imagined, a common upbringing and experience from that other island people at the opposite end of the Eurasian landmass.

Space does not permit elaboration, but you will perceive very personal reasons for my affinity with Japan; I recommend that each individual contemplating ties with Japan visit first, thus finding his or her own inclinations for wanting to forge a closer relationship. There was a cultural intention to my desire for more experience of Japan as well as a professional, engineering, and scientific significance. Cultural aspects are rather pervasive in Japan (as they, at least once, were in Great Britain), and in my experience, the Japanese hardly expect to associate only in a professional way, but prefer also a more personal and intellectual transaction—if you play golf, so much the better, for example; I don't, but discussions of music, poetry, and so forth were, in my experience, more commonplace in Japan than on my home ground.

This prologue will suggest that when asked by my company, Rockwell International, "Would you like to spend six months with Hitachi," I was predisposed to welcome the opportunity. When asked later how I had felt when the chance arose, I replied that it was as though an actor (I live near Hollywood!) had been offered a part in a Kurosawa movie. Any good actor would give his eyeteeth for such an offer, because he will be playing a western role, which is why he is cast, yet he is in a Japanese movie and must benefit the whole drama. (He is not there to become a Japanese, but to creatively donate his "western-ness.") I welcomed the differences—the greater the better—as exciting opportunities for extending my personal life and interests and as a challenge to my adaptability. I also recognized, of course, that, just as a Kurosawa movie is likely to be "influential," so also Japan's role in the world has now become important to the point that anthropologists, sociologists, and economists are probing the reasons for the innovativeness (by which I mean the production of improved or new articles of cultural or economic value) so evident lately (but actually of long historical precedent).

The Rockwell-Hitachi exchange program was set up as a joint, open-ended, precommercial research trial with one person from each organization exchanging each year. Hitachi Corporation has other programs for visiting scientists and engineers (HIVIPS) and welcomes foreign workers, both at its Central Research Laboratory in Tokyo (Chuken, for Chuo Kenkyusho) and at the original Hitachi Research Laboratory (Nikken, for Hitachi Kenkyusho; *Nik-* is the name of the first Chinese character of Hitachi) in Hitachi City. For unaccompanied visitors to Nikken, such as myself, Hitachi managed an "international guest house" in recognition of the sometime difficulties of housing in Japan. For families, company apartments are available. This is a big advantage in the small-town environment, and it's more like the "real Japan."

At the international guest house, Izumigamori Kokusai Ryo, where Hitachi engineers also reside, male foreigners live in very acceptable single hotel-like rooms. Thus, at buffet-style breakfast and dinner, there is much chance for communication. Amenities include tennis courts, international television, and English-language newspapers, as well as company-sponsored mixers. Foreign families tended to drop by this center for tennis, meals, and other social activities. Foreigners were encouraged to partake in company-related swimming, tennis, and other sporting and cultural events, and most did so. From this perspective, then, the Hitachi Corporation has done much to alleviate the sometimes, by western standards, poorer living conditions that may be encountered in Japan. Hitachi encourages learning the Japanese language through videotapes, classes, and other methods.

At Hitachi there have been more visitors from Europe thus far than from the United States; I suspect that Europeans' desire to spend time in Japan is greater than Americans', the latter, unfortunately, being perhaps less well prepared by their relatively parochial education. Educated Europeans expect to have to speak at least one language other than their own and thereby become aware of cultural diversity. Americans must expect that cooperation between Japan and Europe (next-door neighbor Russia, too, is a fascinating prospect!) will go on whether or not the United States itself is also committed.

By 1989, the then new field of high-temperature superconductors (HTSC) was just two years old; since Hitachi and Rockwell (and indeed, it seemed at the time, almost all corporations) were highly interested in this very precommercial development, it was ideal for an exchange subject, with surprisingly minimal legalistic constraints; at the time of this writing the fourth exchange of one person from each company has occurred for research in this field and several more in other fields.

The Chodendo (superconductivity) Center of Nikken was founded partly in the spirit of a longer-term, curiosity viewpoint, as well as (hopefully) of near-term profit. The surprising, unexpected scale of, or uses of, a product in the marketplace are a more common phenomenon in the history of technology than the advocates of presumptuous, so-called market analysis and planning may propagandize. The attendant risk is moderated by starting small, "learning by doing," and gaining the know-how for later unusual developments that become inevitable if enough seeds have been planted. (In the United States, to some extent, this role is played by the small, entrepreneurial, venture-capital companies.)

Although I had already met the director of the Hitachi Superconductivity Center, Shin-Pei Matsuda, a few months before at a delightful shipboard dinner party in San Diego Bay as part of a Materials Research Society (MRS) meeting, we had not predetermined the specifics of the research. I presumed only that Hitachi personnel had looked at some of my publications, and I likewise had reviewed the Superconductivity Center literature. In spite of my earlier comments on the need for cultural and, at least, some language interests, I recognized that the technical program must take priority; if that went well, then other matters would, with a little effort, take care of themselves. With poor research accomplishment, the rest would be for naught.

Upon arrival (which was accompanied by unexcelled Japanese hospitality), I gave a seminar of my recent work and was asked what I wanted to do. I suggested that the Hitachi engineers, too, present some of their ideas, so that I could see if we could forge a synthesis of our strengths. At this point I should say that the manner of interactions of this kind will greatly vary with age and experience; Japanese can be expected to react appropriately to seniority (as described repeatedly in this book—I was 53 at the time). Whereas I was expected to lead a joint endeavor (naturally with due regard to Japanese "consensus-seeking" methods), a younger person might expect to be assigned a more narrowly defined role.

Especially in the latter case, it is enormously important that the right manager be found for the foreign worker. Preferably he or she would speak fairly good English (and, more important, enjoy doing it), be culturally cosmopolitan, expect to learn a lot, and be committed to promoting international interactions long term. The system especially needs to be structured so that such a manager also benefits from the transaction rather than feel that this visitor has been foisted upon him or her.

I was a little surprised that it seemed that we would embark on a project very quickly—I had expected the consensus process and diffusion of ideas to take at least some weeks. In this regard, perhaps the Hitachi Superconductivity Center was unusual; it was newly created in the wave of excitement surrounding HTSC, containing more aggressive, fast-moving, and young individuals, many with experience in the United States and Europe, and I felt that they knew what to expect from a westerner. And so, while they were apparently, I thought, adopting a more western mode for me, I was trying to fit a perceived Japanese mode (exhuming parts of my long-lost British upbringing?). In any event, after only two days or so, we arrived at a general plan.

My own personal background is in the phenomenology of complex ceramics, including synthesis, sintering, grain growth, and related matters. Ceramic engineering has been a particular focus of interest in Japan within the broader materials realm epitomized by the widely admired Kyocera Company. The noted Japanese success in this field is attributable to an emphasis on processing, with more heuristic, chemical approaches rather than the more physical, analytical American stance. Hitachi has a strength in the technology of silver-encapsulated HTSC tapes, and a peculiarity had been noted in that the silver was surprisingly affecting the behavior of the HTSC in a generally beneficial way. How could we understand the reasons for this? Bearing in mind that I had only six months of initial interaction with the group (although I hoped our association would continue later, and it has), I tried to think of an efficient method of studying this bypassing the hundreds of experiments normally required in a multiparameter ceramic system with up to seven elements and extreme sensitivity to temperature and time, for instance.

The particular outcome was that we realized that by placing the unfired 14-cm tape in a constant thermal gradient and afterward examining the lengthwise-polished section in various ways, we could perform essentially 100 temperature runs with each experiment. The synergy gained by combining a Japanese strength, in this case using the long tapes that had become readily available, with the scientific insight of the thermal gradient method pleased everyone. The details of the work, presented in Japan and in the United States, were quickly published.[2]

"Basic" (curiosity-driven) science is well known to be inspired by technology (for example, thermodynamics by steam engines or astronomy by telescopes—originally for commercial shipping and war—remember!), and the progression is something that the Japanese can now expect increasingly to be able to make. The Japanese are looking into the western ideal of leapfrog invention, while the west increasingly scrutinizes, with some trepidation apparently, the successful *dan-dan* (step-by-step) or *kaizen* (improvement, implying many small, never-ending steps) approach of Japan. Much will be achieved by an eventual synthesis of these creative viewpoints.

At the Superconductivity Center, the open-office system, discussed elsewhere in this book, was paralleled by an open-laboratory approach (refreshing, compared with the oft locked-door fiefdoms of laboratories in the west). Each piece of equipment has a "keeper"; the younger personnel "learn by doing" and "start small," training on the job with the senior personnel, in the well-known Japanese way. Few people enter the system as Ph.D.s; some that do have foreign Ph.D.s. Generally the young people enter straight from the bachelor's level expecting to learn on the job and can be molded in the company "family image" as they anticipate, of course, lifetime tenured employment. This does not mean that a lifetime of research is necessarily anticipated; different roles may be "agreed upon" and expected as an individual moves through the corporation. Doctorates may be acquired later, as D.Sc.s are in Europe, by associating with a university and using the body of research publications. One now understands how the "club" interests at universities predispose and aid students for this later group behavior (see Chapter 9).

This sociotechnical difference between cultures leads naturally to a different outlook on research. In the west, the Ph.D., who controls most exploratory research, has already been groomed for seeking the big, individual, original, conceptually unique, historic "leap forward" as a goal. Not so in Japan, where group harmony will be best preserved by many agreed-upon, even allocated,

[2] P. E. D. Morgan, M. Okada, T. Matsumoto, and A. Soeta, "A Thermal Gradient Technique for Accelerated Testing of T1-HTSC (or, for That Matter, Any Ceramic)," in T. Ishiguro and K. Kajimura (eds.), *Advances in Superconductivity II: Proc. 2nd. Int. Symp. on Superconductivity, ISS '89*, Tsukuba, Japan, November 1989, Springer-Verlag, Tokyo, pp. 435–38).

small steps of achievement, integrating sometimes to a greater end, albeit not with the fireworks, excitement, and Nobel Prizes of the western model. Japanese often comment that they do not distinguish between science and technology as they hear westerners discuss them. This could be the subject of another article, but the process of science seems to be viewed as essentially the same as that of technology. For better or worse, I have yet to meet a Japanese person who has read Karl Popper's work, that of his acolytes, or the later scientists such as Kuhn, Polanyi, Medawar, and Hallam, and others who have discussed "philosophy" of science from a working scientist's social viewpoint (and whose books I recommend).

Few working scientists or engineers in the United States have done this either, let alone indulged in Buddhist readings, but it is the sort of interest that the western scientist can bring to Japan, since creativity in general is a fascination of the Japanese. Since it is an active interest of mine, it was a credential that I could bring into play, and therefore, in an English conversation group I was leading at Hitachi, we discussed scientific creativity (indeed, any creativity) as well as many other topics. (Occasionally I was guilty of "lecturing"— not a good idea.) This is an example of how a foreigner's special strengths may be appreciated by the group in a Japanese corporation.

In my personal interactions I did run into a few situations that were not unexpected, after the reading I had done, but which are worth passing on here. The Japanese don't like to say no; "it is difficult" or "maybe" usually means "no." It is much better to ask for suggestions than make requests; demands are out of the question. If a question or request is not answered, don't push, but wait and rephrase the inquiry, give reasons for the request, even wait for a social situation (over sake, for instance) gently to raise the issue. Humor, if you can bring it off, is acceptable, even helpful, to soften a position. Sometimes, in trying to get an answer to a "difficult" question, you may feel that you have been given the answer to a completely different or unasked question. Let it pass. Assume that language difficulties may have caused this; even if they haven't, this allows a way out. Learn to think of this from the Japanese point of view and always allow time.

Another issue is the *uchi-soto* viewpoint—roughly, but more strongly, "us and them," in English. If, according to plan, you integrate well into your group, becoming uchi, you can become soto to another, competing group. Unofficial attempts to build bridges, without the aid of higher authority, say, from the company above, and especially without proper introduction or advance warning, can be viewed with extreme suspicion. A person may claim to be too busy to see you, for example. This is understood to be a put-off, and you would have to devise a plan to ameliorate the situation for the future. This may also occur between you and people in other companies. It may be that you know a person in one group very well but another person you want to meet is in a competing group and, although this second person may share your larger professional interests very closely, nevertheless, uchi-soto may be an overriding factor; treading on toes is to be avoided at all costs. I would not over-worry about these situations to the point of paralysis, however. Individual temperaments vary much in Japan, as elsewhere.

Once again, this can only be a personal recapitulation of my own experiences, but I hope I may emphasize that while you learn (as from other chapters of this book) about Japanese ways, corporate culture, and so forth, you should not become overly inhibited. In the end you must be yourself cast into this role, use your own inner resources in the part so that the production is improved; remember that sincerity, giving, understanding, concern, patience, sensitivity, and warmth are truly international attributes that will be appreciated and reciprocated. Speak with actions, for the Japanese are ruled first by human relationships.

It is occasionally said that the Japanese are ruled by emotions more than westerners (a surprise to those who think that the Japanese are taught to suppress emotion—yes and no!). A gross generalization would be that, whereas many Americans regard their work as a means to buy the pleasures of "real" life when they are not at work, the Japanese see their work as nearer to the core of what they are—akin to the way in which more creative people everywhere generally see themselves. Perhaps westerners would do well to wonder if the sensibilities of music, poetry, literature, and religion, which we hope illuminate the curiosity of our lives, are not nearer to the surface in Japan (as I believe they are in Europe also). Wonder why the Japanese observe *sakura* (cherry blossom time) with a kind of melancholy; for life, like the beauty of cherry blossoms, is ephemeral (perhaps even an insubstantial dream). Know a little of the differences between the Buddhist and Judeo-Christian views of the world and attempt to probe the more than 2000-year-old Japanese soul or spirit (*kokoro*—a word habitually used in literature), and you will be on the cultural track at least. All of this will give you plenty to do, along with learning some language, while you have those slow start-up periods that are often commented upon in Japan.

Many small, unanticipated, pleasures occurred during my stay; for instance, when two of my colleagues from Rockwell, who had not been to Japan before, came visiting, I was able, with the knowledge I had acquired of the local region through the hospitality of my Japanese hosts, to accompany the visitors to the local gardens and temples. This was a welcome respite from business, and an opportunity to show off some places unlikely to be visited normally, say, by tourists. In this way I was able to convey, I hope, some of my own interest in and affection for things Japanese.

On returning (unless, like Lafcadio Hearn, you never return) your experience will be of value in ways difficult to anticipate. One of my personal pleasures is to escort Japanese visitors to the United States to places and events that I now better know they will greatly appreciate and to enjoy the Japanese sense of humor that emerges quite quickly if it is cherished and reciprocated.

After I returned to the United States, I revisited Japan two or three times a year for business and pleasure, now with added enjoyment resulting from the earlier experience, from the humorous and anecdotal stories I had heard, and from the knowledge of the language and customs I had gained during my sojourn.

Later Work Reflections

After I first wrote this chapter, my Hitachi-learned experience led, unexpectedly about seven years later (1997) and after many more visits to Japan, to an invitation to be a guest professor for three months at Osaka University ("Handai," for short), where I had a new responsibility with students working for their M.S. or Ph.D. degrees. In this case, the students may have perceived that they, too, were taking quite a risk in working with a westerner. It was here that my earlier experiences helped again in terms of allowing longer time constants for developments than might be expected where language and cultural differences are not present. Patience is extremely important, of course. However, if students sense that one "knows" Japan, then they will be more confident; they hope anyway to benefit from the western "actor-scientist" and can be expected to rise to the occasion if properly encouraged.

The same rules, demanding that good technical work and publications come first, apply (that is the output that our public sees corresponding to the finished movie). However, I was able again to give a short course on "Creativity-History, Philosophy of Science, etc." in the form of "soft" seminars designed to be enjoyed, by doing exemplifying puzzles and so on.

Later (1999), I did a one-month stint at the NIRIN government lab, in Nagoya. Returning to an earlier theme during this visit, I was again asked to give a "soft" seminar (quite apart from the many "hard" seminars I gave), so I replied, "I am an American and I am different, *so the soft seminar must be with beer and pizza.*" It was overwhelmingly embraced as a good idea, as you might imagine. A Russian puzzle, needing a heuristic guess (or moment of epiphany) and with a mild sexual innuendo, after some beer, was also amusingly tackled.

With these added experiences I dare to add a few more personal ideas that might be useful.

It might be quite unwise to accept a job in Japan without several preparatory steps. It would be good to discuss the nature of the work in some detail before committing. This should not be taken as negative by the company, even though they would not expect to do so with a Japanese hire. Westerners are supposed to be more individualistic and, with the growing awareness in the Japanese companies that some of the foreigners are not happy, it should be seen to be a plus that the applicant is suitably cautious. The applicant should be selling his *controlled* individualism and creativity that will *harmoniously* benefit the group. It would be even better to travel to Japan to prospect the job. The company probably would be surprised at the idea but should appreciate the initiative (this assumes that the company doesn't plan an interview; I have never heard of it being suggested). I realize that this would seem to be expensive, but there are creative ways nowadays of traveling quite cheaply, and young people often know how or can find out; it would be another sign of initiative. I know one chap who traveled to Japan from England *by rail!* Yes, he took the Trans-Siberian Railway all across Russia and a boat from Vladivostok to Yokohama. At that time, in 1989, that route was incredibly cheap, but only if the tickets were purchased in Russia or Eastern Europe and not before leaving England. Anyway, we were impressed.

A foreigner working in Japan will do so usually for not more than two or three years; the foreigner's later success must therefore depend upon work designed for the wider-world-peer-group/publication/résumé audience, but this need not conflict with company goals. It mandates a different work policy for the foreigner as against the Japanese, who will work strictly within the expectations of the company culture, with promotion most likely within management if successful. The entering Japanese, as discussed elsewhere, will most often have a bachelor's degree only and is to be molded by the company on the job; he will anticipate being employed from college to retirement with the big company, possibly occasionally with an authorized move to a keiretsu (conglomerate family) related company.

The visitor most often will arrive with some advanced training beyond the bachelor's level, often a Ph.D., and, it is assumed, with more aptitude for advanced work. (In fact, there have been difficulties when the arriving foreigner is younger with only bachelor's-level training.) It is preferable that the companies take advantage of this greater experience by allowing the foreigner more flexibility and leeway to work with a wider group and certainly to influence his or her own group in a way suitable to an increased capability. However, there are many sad stories where the company gave the visitor simple technician-type work similar to the normal established company policy for "climbing up through the ranks." Both the visitor *and the company* lose thereby. If the visitor and the company each understand the visitor's goal, both should be able to benefit especially as the longer-term goal should be rewarding *jinmyaku* (networking).

The situation of spouses accompanying partners to Japan is interesting; certainly a commitment by the spouse to the same positive, somewhat cerebral, but perhaps emotional goals is an advantage; just tagging along is likely to be a disaster. There are many cases of unhappiness among spouses, but the really investigative, outgoing, inquisitive, and enterprising ones I have heard about seem to have had a very good time. In a general way everyone should discount (anticipate) their future feelings (in the sense of the stock-market mechanism of discounting) so that inevitable disappointments, homesickness, etc., are built into the expectation.

Much is written about "consensus" *(goui)* in Japan. Often it seems to me that the word is not used in the sense that a westerner immediately understands it, which is as consensus of opinion. Much more often it has to do with consensus for *action*. In a football team, the plays are decided largely by the coach(es), there may be little consensus of opinion with the players, all plays are not exhaustively discussed and agreed upon by all, and maybe some players disagree with the coach and each other; but during the game, there must be consensus of action for the overall winning strategy. Thus we have both consensus of action and suppression of individual desires/opinions to achieve the greater goal at the appropriate time. Similarly we may say that, when actors perform a play on the stage, the director has decided much previously on the action and will be not be utilizing a complete "opinion consensus" of all concerned but will engage in more or less discussion. When the curtain goes up, consensus of action has been achieved. The Japanese consensus seems to me to be like these examples. Individually Japanese workers will tell of disagreements but necessary sublimation of individual desires for this seeming winning unanimity.

The Japanese garden provides a good analogy for the attention to detail that the Japanese so highly value. To suggest that the garden be "natural" or "easy to take care of" or "low maintenance," all western ideas, makes no sense. It must be the impressed values of the human designer/artist/gardener upon nature. It must require meticulous, never-ending attention against runaway growth, fallen leaves, the elements, seasons, and so forth, and reflect heightened nature through the seasons as an artist might see it as a painting or in a poem. "It is never finished" and can never be good enough.

In discussing basic science research in Japan, I have often cited the work at Kamiokande as the very best example of "really pure" research. Kamiokande is a an old deep mine (to limit cosmic ray interference) in Japan where experiments were set up to detect the possible decay of the proton, as predicted might occur by some of the theories of high-energy physics and origin of the universe (part of testing a "Theory of Everything"!). Not only does no one know any possible use of such erudition, but no one can presently conceive that any use could ever materialize. So far no decays have been seen. That is certainly "basic science," but it had another distinctive basic characteristic in that something entirely different was discovered serendipitously (that is, an a priori unsuspected observation was enabled that no one was looking for). This was the detection of neutrinos liberated by a supernova in 1986 in one of the "nearby" stars in the Magellanic clouds, the most conveniently observable supernova in about 500 years. The observations were critical tests of previous theoretical models about star deaths. Even less do, for example, NEC and Hitachi imagine that they will ever harness the energy of supernovas! A telling comment on this is that few Japanese scientists to whom I have mentioned this appear to have heard of it; this obviously does not include high-energy physicists, whom I don't generally personally encounter, but it does include many university and industrial scientists. I have concluded by this and other means that most Japanese scientists are narrowly focused on their immediate disciplines (as are many western counterparts) and not on general science as an "outlook on life." I do, however, offer this as a counterpoint to those who imply that Japanese science is only applied.

I attend many scientific meetings in Japan and I have noticed a curious fact: many older (definitely over 50 or 55 years old) and learned foreign speakers, who have usually spent their whole lives researching their disciplines, are invited or otherwise attend. They often guide the thinking at these meetings. While they are expectedly somewhat more conservative and cautious, they complement the also necessary, impulsive, energetic, and even radical young foreign people. With the accumulated experience that comes with age, they can deliberate on and contrast between "good-crazy" and "bad-crazy" new ideas. These meetings tend to lack Japanese equivalents giving talks. Where the experience of age is so valued as it is in Japan (for example, the artists and craftsmen who are declared "national treasures"), we have to wonder why this appears not to be so in science—I have many ideas on this—but this is (and probably will be) another article.

Several people I know who have been to Japan have commented that Japanese reserve is skin-deep; underneath resides a quite emotional, sentimental nature. The most highly regarded early Meiji-era foreign visitors seem to have formed an emotional attachment/relationship with Japan and its people on this level, and this was reciprocated. Thus the attachments made by Lafcadio Hearn, Leroy Janes—"An American Samurai"—(each in quite different ways), and others are remembered with great fondness. I believe it is a little known fact that Langdon Warner, who advised President Roosevelt against bombing Nara and Kyoto in World War II, is buried at Horyuji, one of the most famous and oldest Japanese temples. Some of these men seem quite unlikely to have left much of a mark on their own societies, and in fact some had prickly personalities that caused abrasions in their original worlds (and, indeed, in Japan, but that has been somewhat edited out!). The emotional bridges seem to have counted for much more than any downsides due to wide cultural gaps.

Most foreigners know that Japanese will commonly say "Japan is different" or "Japanese are different"—that is fine—but try, only when you can be humorous in the right situation, saying "Americans are different." I have heard that the Japanese have no cultural equivalent for "many a truth is spoken in jest," but, nevertheless, truths can be gently conveyed in jests. (Interestingly, I have heard that in Japan, perhaps, "many a truth can be conveyed in poetry"—an interesting idea, shared in fact with western poetry.) Japanese, like most people, relish humor related to absurdity, so, if you do dare to jest in Japanese, you have the possibility of making it quite absurd precisely because it comes from a foreigner—as a "weary" observation from an older foreigner on the "fecklessness" of the young" for example. It will indeed be funny as it pushes universal buttons. Older Japanese have referred to the young as "soggy noodles," for example—I expect the young to respond in style. The other worldwide butt of humor, sex, is probably best done in English—"This is a joke from America"—and after sake, etc.; and certainly requires the usual cautions.

I have heard much about the fact that many visiting scientists and engineers who work awhile in Japan find that the experience gained is not valued later when they come to return to their own countries (this is mentioned elsewhere in this book). Partly this is because the individuals themselves overrated the benefits that others would perceive. It is certainly necessary with all experience to sell the results aggressively and not expect the experience to speak for itself. It's what the individual himself makes of the doing that counts, not the act of doing itself. So look ahead continuously and expect to answer the question, "In what ways did you benefit from your experience in Japan that can now help our company?" Of course, the question will likely be less bluntly and more circuitously posed, but that is what any potential employer really wants to know.

Human nature being what it is, it is inevitable that problems for visitors to Japan, generated through their own mistakes or lack of enterprise, are, perhaps, going to be blamed on the Japanese system. Japanese will, I expect, discount this.

A CIVIL ENGINEER'S EXPERIENCE WITH THE JAPANESE CONSTRUCTION INDUSTRY

Charles R. Heidengren

In the present economic environment, Japanese firms, as well as most international organizations, have realized the importance of working with foreign-born engineers. Not only does this policy enhance their ability to work more efficiently in an environment outside their native land, but it also promotes meaningful exchange of technological ideas and information.

My comments are based on more than 10 years of living and working in the Japanese construction industry. Five years were spent with a large private civil engineering consulting firm, and $5^1/_2$ years with the engineering and construction division of a large steel manufacturer, where activities included turnkey design-construction projects and general contracting. My activities included business development, technical proposal preparation, prequalification, competitive bidding, and contract administration. I was hired as a specialist.

The construction industry in Japan, like that in the United States, plays a crucial role by providing the structures that house and facilitate virtually all other economic and social activity. In addition, this industry has historically played an important role outside Japan, not only through direct export of goods and services, but also through exercising leadership in opening opportunities for other Japanese businesses and for intellectual exchange that improves international understanding.

The importance of technological leadership is widely recognized in the industry as a key component for enhancement of the competitive position in an increasingly global marketplace. In order to remain competitive in international markets, traditional technological advantages must be maintained and the skills needed for competition must also be developed. It is in this latter area where the foreign-born engineer can play a key role in Japanese industry.

Civil engineering technology in Japan is excellent. Management-related technology is, however, somewhat lacking. In my experience the organization of project team activities, scheduling, and cost control in Japan were not always as vigorous as "western" practices. Computer technology, too, was at one time limited by a lack of suitable software in many disciplines, including construction engineering; this situation has now improved. Japanese civil engineers practicing in private industry tend to rely on their own personal experience or on government manuals.

Specifically, as a foreign-born engineer, one can expect to observe the following deficiencies in engineering practice:

▶ Lack of use of innovative technology by private consulting engineers. More creativity is shown, however, by the large contractors, suppliers, and manufacturers.

▶ Japanese industrial and other standards are used for design criteria, materials specification, and the like even when not the most suitable for overseas locations, because of climatic and topographic conditions, such as in very hot or wet tropical areas.

▶ Management policies are not always very efficient, thus leading to lack of coordination and poor organization of work methods and procedures. Inadequate staff scheduling often results in cost overruns and low profit margin.

▶ Design or contract drawings are sometimes poorly laid out and inadequate in detail.

▶ Most engineers in Japan are "generalists." State-of-the-art technological advances may be limited because few specialists are available to perform geotechnical, hydrologic, environmental, or structural work.

▶ Higher costs to the owner may result from a conservative approach to design without innovation or creativity. An ad hoc philosophy leads to setting (and trying to meet) impossible deadlines. There is no time for innovation or creativity; instead, designers rely on design manuals from the government or manufacturers.

What is Positive

At the top of the list of strengths are the excellent Japanese work ethic and sophisticated designs based on advanced design criteria. Japanese technology in tunneling and earthquake engineering is on a level with the most sophisticated and best in the world.

Contractors and designers in Japan maintain realistic factors of safety in their work. Corners are not cut to save money. Basic research is very often implemented for complex engineering problems, particularly those involving new technology.

Simulated testing, model testing, and computer analysis have accomplished a great deal in applied research leading to development. This work is most often done by manufacturers, contractors, and suppliers.

Japanese company employee training programs are generally good. A foreign engineer can generally expect to be involved in such programs.

Diversity

In Japan, construction investment equals about 16 percent of the gross national product. This compares with only about 10 percent in the United States and other industrialized countries. The previously underdeveloped state of infrastructure in Japan, including parks, highways, bridges, airports, sewer systems, and housing, has accounted for most of this difference. In most western industrialized countries including the United States, construction companies usually function only as builders, whereas in Japan construction firms are involved in architectural design, urban planning, civil engineering, and land development. They are also now at a turning point in their attitudes toward marketing of services.

Foreign-born engineers will observe another important difference between practice in Japan and that in the west. There is a high level of research and development by Japanese general contractors. Large numbers of top researchers working in laboratories furnished with the latest scientific equipment comparable to that found in American universities are working on problems related to new markets. Some of the topics being investigated with keen interest include advanced technologies in biochemistry, space development, air pollution, and hazardous and toxic waste disposal.

Foreign-born engineers will find the attitudes, motivation, and capabilities of young civil engineers in Japan to be different from their own, generally speaking. Few seem interested in overseas assignments, because of a preference for the Japanese lifestyle and family goals, and because of problems they would have in overseas countries with two-career households, children's education, and communication. Communication in a second language, both written and oral, is very difficult for many Japanese, not only engineers. Efforts are being made in both the public and private sectors to solve this problem with intensive foreign-language studies.

Most contractors and consulting engineers in the United States encourage young engineers to participate in the activities of professional societies such as ASCE (American Society of Civil Engineers), ASME, and AIA (American Institute of Architects). Professional society meetings, conferences, and continuing education programs are normally considered an important part of career development in western countries. The average Japanese engineer does *not* participate to any great extent in professional activities. Instead, most efforts are devoted to the company's goals.

Future Expectations

In the future, I expect that Japanese practicing engineers will participate more actively in international conferences by presenting technical papers and working on committees. They will be more willing to adopt ideas and developments from the United States, the United Kingdom, France, Germany, and other countries.

Professional attitudes and understanding will also be very important to carrying out the "megaprojects" being initiated in Japan. About $15.8 trillion yen was spent in 1996 to improve, maintain, and repair the Japanese infrastructure and this amount is expected to grow to 21.7 trillion yen in 2010. Professionalism will play a significant role in satisfying the growing need for cooperation between Japanese and foreign engineers.

In the future, there will be more investigation and awareness of differences between Japan and the developing world in such matters as labor, materials, equipment capabilities, environmental conditions, and sociological pressures. There will also be more willingness to share Japanese technological innovations and developments with engineers everywhere.

A EUROPEAN ENGINEER'S VIEW OF INDUSTRY
Dimitrios C. Xyloyiannis

This case study stems from my intermittent involvement with the Japanese people since 1963.

My first encounter with the Japanese was in 1963–64, during repair of a tanker ship at the Hitachi Shipbuilding Company. Later, in 1974–77, I supervised (as a surveyor, just after getting my M.S. in mechanical engineering) the building of 12 bulk-carrier vessels with a capacity of 33,000 deadweight tons at the Kanasashi Shipbuilding Company. During that time, I had to travel extensively to many places in Japan for the shop trials of the ships' main and auxiliary machines.

For the next 12 years I worked as chief engineer for a multinational pharmaceutical company. In 1989 I was transferred to my company's Japanese affiliate as project engineer for the construction of a factory in Saitama and a research institute in Tsukuba.

My experience has taught me that when dealing with the Japanese one has a high probability of working out a good relationship if one considers a Japanese partner to be trustworthy and of at least the same capabilities as oneself. In most cases, I have found that when there is a conflict between a foreigner and a Japanese, the Japanese is right most of the time. Of course, it takes two people to make a conflict, and I do not claim that the Japanese are always right, but at least they are often responsible to a lesser degree. Therefore, we foreign engineers have to look into the cause of any conflict that arises, see where we are wrong, and admit it immediately.

Foreign engineers who plan to work with the Japanese, especially in Japan, must do their homework before they get involved in any Japanese business. I would suggest they they learn as much as possible about Japan and its people: its culture, customs, methods of decision making and follow-through, business values, methodology, and working hours. Socializing with the Japanese and studying their behavior can also enchance business relationships.

I personally found the study of the Japanese language very beneficial. I did not have to master the language, but by trying to learn it I could understand much more easily why my Japanese partners have so much difficulty understanding foreigners.

Making trips to the countryside (small cities such as Fukui and Fukushima) helped me to observe the real Japanese way of life and gave me some insight into the people's behavior. Big-city society is somehow "contaminated," as a Japanese colleague once expressed it to me.

To paraphrase a familiar saying, when in Japan, do as the Japanese do. My experience is that by applying constructive behavior, one has a much greater chance of being successful. Constructive behavior includes everything that will contribute to gains for all parties involved. In Japan the sense of balance is very intense. Therefore, deals and bargains struck on a win-win basis (not I win, you lose) offer the highest possibility for success. In a nutshell, it is wise to do the following:

- ▶ Encourage your Japanese colleagues to do most of the talking.
- ▶ Listen carefully and verify understanding.
- ▶ Avoid comparing cultures (especially criticizing) unless you see some positive similarities. Japanese culture has survived and been successful for centuries, just as it is. By criticizing, one loses ground and to a high degree creates, if not a negative, at least an uncomfortable environment. I would like to emphasize this point, because I have seen many foreign engineers fall into this trap.
- ▶ Avoid showing off. The Japanese people are very eager to learn and humbly accept that they have learned and can learn from foreigners, but they dislike being lectured.

▶ Be consistent and be a doer. In other words, be sure that what you say, you can do.

▶ Keep a written record of cases in which you have failed or succeeded, and make guidelines that fit the working environment you are in. What works in my case does not necessarily work in your case.

In conclusion, one should consider it dangerous to take one case or a few cases as rules, no matter how authoritative they seem. This is particularly true in an area as complex and controversial as the relationship between people, from different cultures and sometimes different educational disciplines, engaged in reaching a common goal, especially if they have dissimilar interests.

References

I have read many books on the subject of Japanese business relationships, but here is a list of books I would thoroughly recommend:

Mark Zimmerman, *How to Do Business with the Japanese: A Strategy for Success,* Random House, New York, 1985. Zimmerman presents excellent suggestions for working successfully in a Japanese company (Chap. 17, p. 254).

H. Jung, *How to Do Business with the Japanese,* Japan Times, Tokyo, 1986.

Takeo Doi, *The Anatomy of Dependence,* translated by John Bester, Kodansha International, New York, 1982.

A. Young, *The Sogo Shosha,* Charles Tuttle, Tokyo, 1989.

21

Observations on Working in a National Institute and a Research Laboratory in Industry in Japan

Joyce Yamamoto

INTRODUCTION

In preparing for this article, I had to sit down and organize many thoughts, ideas, and experiences. I will concentrate on working as a researcher in both a national laboratory and a corporate research and development (R&D) laboratory. When I was asked to write about my experiences of working in the field of R&D in Japan, I was immediately aware of the potential to increase understanding on the part of the host researcher of some of the difficulties a visiting research scientist may encounter. As increasing numbers of researchers, both from industry and academia, come to Japan, the need for understanding each other's problems and how to deal with them will increase. What I will try to do is to objectively describe in general terms some problems and impressions that I have encountered working in both types of research environments.

Being a Japanese-American female scientist with a rudimentary grasp of the Japanese language presented other problems and experiences. Because of cultural and social differences, the position of women in the various scientific fields in Japan is not a prominent one. As Japan and Japanese science continue on the path to internationalization, contact with more people who do not quite fit the popular image of a *gaijin* will increase. I imagine that I represent an extreme case, but I hope that my input and descriptions will increase awareness of the experiences that the visiting researcher may encounter.

This article is broken into three general parts:

1. Experiences and descriptions of working in R&D
2. Descriptions of various difficulties and suggestions for the host in order to ease some of the problems that I encountered
3. My experiences of being a woman scientist working and interacting with Japanese scientists

Since my situation is a little out of the ordinary, it is my goal to describe my experiences and to offer some suggestions or hints to help future scientists make their stay in Japan fruitful and enjoyable for themselves and their hosts.

WORKING IN A RESEARCH LABORATORY

In talking with other foreigners, I found that my adjustment to daily life in Japan is not different; using the train or bus system, adjusting to Japanese cuisine, and trying to communicate my thoughts using a handful of Japanese words are challenges common to most visitors. In starting a new position, there are always difficulties in learning how to function in the new environment. The same is true in R&D. In many ways research is similar to other professions, but there is a "second work environment," or the lab, which may require additional time and effort to master. Therefore the demands on the hosts' time and effort are increased to accommodate the transition to the laboratory environment. Of course, not all research is done in the laboratory, and there are different situations depending on the nature of the research. The laboratory may vary from a personal computer to a complex high-power linear accelerator.

Working in the national laboratory was very much like working in a university. I found the atmosphere of the national laboratory to be very relaxed and cordial. Discussions on various topics were encouraged, and the flow of ideas was always interesting. The best surprise was that the language barrier was minimal. Many of the researchers are encouraged to study overseas and had spent an extended period of time outside Japan. As a consequence their knowledge and command of English made communication very easy. They were also aware of the differences in conversation styles and were able to interact in extended discussions, resulting in efficient and fruitful conversations. This also translated into a reduced training period. Instructions on how to operate instruments were quickly and easily communicated with a minimum of confusion. I was pleasantly surprised at the rapid pace at which I could start and run my own experiments. The one problem I encountered was that the control panel buttons on many instruments were written in Japanese. My solution was to write the English translation of each button, knob, or lever next to the Japanese label, making the instrument bilingual. This had the added benefit of increasing my technical Japanese vocabulary.

Working in a corporate R&D laboratory was very different and presented many new situations. Since it is a private company, the end goal—the product—is heavily stressed. The researchers' goals were well and narrowly defined. I was amazed at how well all the members knew their roles in the project, forming a well-organized, efficiently performing team. I learned much by observing the way the group functioned to achieve its goals. The group members were very supportive of each other's experiments. Although my supervisor had a good command of English, it was usually the case that the technical staff did not speak English, and this sometimes caused misunderstandings. Until I established a working vocabulary with each person, I frequently needed assistance. There were also a few more new rules that I had to add into my daily routine. For example, I had to get used to wearing different slippers while in different laboratories, and there were more forms to fill out (in Japanese), which took a while to do until I learned which kanji were appropriate and the correct location for each one.

PROBLEMS OF THE VISITING RESEARCHER

The visiting scientist who has come or is thinking of coming to Japan to study has weighed the advantages and disadvantages of dealing with the language and cultural differences. Since Japan is gaining a wide reputation for its advances in technology and research, it is not hard to see why scientists feel that it is necessary to establish contact with their Japanese counterparts. The experience that most scientists hope to gain is knowledge of the state of their field of research in Japan, to increase their awareness and appreciation of the Japanese culture, and to make new friends and acquaintances.

To better introduce this topic, I will try to describe what a visiting researcher may be feeling, the anxieties and apprehensions, to give the reader an idea of the researcher's mind set. (Another resource to consult concerning this topic is found at the end of this chapter.) The language barrier is by far the largest problem and the most difficult to bridge. It's very difficult to estimate the size and magnitude of the problem, prior to actually being put into this situation, and how it will affect the researcher's productivity. From the visiting scientist's point of view, especially one who has no or little knowledge of the Japanese language or culture, this can be a frustrating experience. The visiting scientist will be accustomed to communicating his or her thoughts and ideas efficiently and functioning independently. Being able to communicate both orally and in written form is essential to the visiting scientist, and therefore it is a difficult transition to make from self-sufficiency to dependency on others for the simplest of tasks. The visiting scientist will undoubtedly be very uncomfortable and feel very isolated because this is such a new situation.

The communication problem has also been addressed by other scientists, and there are different points of view on whether it is worth taking the time away from research to study Japanese. Since journals and meetings are conducted in English, and since the level of English proficiency is high among Japanese scientists, it is one view that the time is better spent focused on research.[1] To really develop an appreciation for the Japanese culture and people, an understanding of the Japanese language is needed. Therefore some command of Japanese, I think, should be attained, if only to function in life outside of the laboratory. Ultimately, it is a personal decision, to weigh the pros and cons of making the commitment of learning the language and how it will benefit the researcher.

The visiting scientist will probably have many questions before arriving in Japan concerning both daily life and the work environment. In addition to the many questions about the language barrier, there will be questions concerning the work atmosphere and style. For example,

1. How will the visiting scientist be accepted by fellow group members?
2. To what extent will the visitor's personal work style be tolerated or encouraged?
3. How will the visiting scientist fit into the general day-to-day workings of the group? Will the visitor be left to work on his or her own or work closely with one person?

---

[1] Gerrit E. W. Bauer, "Can and Should a Physicist Learn Japanese?" *Oyo Butsuri,* Vol. 60, No. 8, 1991, p. 813.

The advantage of working closely with one person is that it is an efficient way to become familar with the laboratory and the researcher, but it is a large time commitment for the Japanese host. This also tends to inhibit interaction with the other group members. Working in a group situation tends to force interaction with all group members, but it may take longer to learn how to function in the lab. There are many different situations, and they will have to be evaluated depending upon the field of study, the nature of the work, and the willingness of the research group to integrate the new visiting scientist.

What makes working in R&D different from other professions? There are many similarities in terms of the skills needed to function in the workplace, and the many suggestions that the other authors have presented in this book certainly apply here. The unique problems associated with working in a laboratory set it apart from the office environment.

Care needs to be taken to ensure that proper operation of equipment or instrumentation is understood. Usually the visiting scientist will be familiar with the general principles involved in operating an instrument. In using Japanese computers, I found computer programming to be frustrating, not because of their performance, but because of the language barrier. Even though the program language itself was written in English, the manuals to use the programming language and computer system were written in Japanese.

When the computer was used to control instrumentation, the consequences of a mistake quickly increased in importance. No one wants to needlessly damage the instrument, wasting time and money for repairs, but pressing the wrong button by mistake is more likely if the visiting scientist is not properly trained on the instrument. Some companies have English translation manuals for their product; having these on hand would help to avoid misunderstandings.

The nature of certain areas of research can be dangerous, and this is one thing that sets the lab apart from other types of work environments. The issue of safety in the laboratory is very important and cannot be overemphasized. Learning to use the equipment safely is important in avoiding damage, but the laboratory can also include many potentially harmful hazards. The researcher many be working with:

1. High or low temperatures and/or pressures
2. High voltages and/or currents
3. Toxic chemicals and substances

4. Ionizing radiation (X rays, gamma rays, etc.)
5. Any combination of the preceding

Knowledge on how to use the instrument and what to do in case there are problems (laboratory safety rules) is very important. In addition, the visiting scientist should also be responsible for learning the important kanji and what to say in Japanese should an accident occur.

This brings us to the next topic. In the unfortunate case where something goes wrong or there is an accident, the researcher must also be aware of the lab emergency procedures. Most labs have regular saftey drills, so the researcher will probably understand the general procedures, but researchers should also understand what to do in all other emergencies.

Another area where the researcher may need additional help is in gaining access to foreign scientific journals. It was very sobering to walk into the library at lab only to see Japanese journals—then exhilarating to find the English journal section. It is important to maintain knowledge of the current work through journal publications, so access to databases, the JISTEC system, and patent search services will help to ensure good-quality research and minimize the feelings of isolation the researcher may experience.

A FEMALE SCIENTIST WORKING IN JAPAN

In Japanese society, it is not common for women to pursue a career in science; therefore, many coworkers may not be accustomed or willing to accept working with a female colleague. Since more women in other countries are choosing to enter this profession, more women scientists may also choose to study in Japan. If the researcher has received a university degree in a technical field, i.e., a B.S., M.S., or Ph.D., and has chosen to work in Japan, her commitment to her career is genuine. Women's training is the same as their male counterparts', and they have progressed through the same evaluation procedures. The female researcher will most likely be very accustomed to working in the male-dominated workplace. She will conduct herself professionally and expect to be received in the same manner. The presence of a foreign female in a research environment is a new phenomenon in Japan, and this situation is where the hosts will probably have to make the largest adjustment.

Being a female scientist, I've found the largest barrier that I've encountered to be interaction with other Japanese scientists. There has not been a problem within the research group, because after the initial period of adjustment, I've been made to feel comfortable and accepted. In attending conferences and making presentations, the reception is quite different, and I was usually kept at a distance. It could be said that my treatment is because I am a foreigner, but I have been at conferences with other foreign male scientists and their reception was much different. The perception of the "minority" scientist by fellow Japanese scientists needs to be improved to strengthen the technological ties between Japan and other countries.

My situation is complicated one step beyond being a female scientist in Japan. I am also of Japanese descent; therefore my outward appearance is identical to a Japanese citizen's. But that is where the similarities end. It is said that a foreigner is easily identified by more than just outward appearance, that his or her actions and attitudes are easy to identify. I have found that even though I can see differences between my actions and those of the Japanese, they do not seem to be recognized by the many Japanese I have met. I can only guess at the reasons for their behavior but I feel that this has been a major disadvantage as non-Japanese looking foreigners are often paid special attentions and cares.

SUMMARY

As Japan tries to increase its contact with other countries and cultures, and especially in technical and scientific areas, where both parties have much to benefit from collaboration, the cultural barriers still are very high. My situation represents an extreme case, and it is likely that other female scientists have not had experiences similar to mine, but it illustrates that increased contact with many different types of foreigners is needed to gain a general level of acceptance for all foreigners. It has to be realized that not all visitors will look foreign and will be male. Greater emphasis needs to be placed not on appearances, very much a Japanese cultural trait, in dealing with gaijin, but in the potential benefits of establishing and nurturing these interactions. The exchange of scientific and cultural information will aid in understanding how different countries can cooperate to increase the internationalization of Japan.

Both the visiting scientist and the Japanese host can learn much from each other to reduce the cultural gap and strengthen international ties.

I think overall working in Japan has been a favorable and worthwhile experience for me. In the future, I hope to maintain the ties I have established in Japan and that by writing this article I can pass something on to benefit future international exchanges.

Reference

The British Chamber of Commerce in Japan, *Gaijin Scientist,* Obun Printing Co., Inc., Tokyo, 1990.

22

Observations on Working in Japanese Universities

Hiroshi Honda, Giuseppe Pezzotti, and Michael W. Barnett

INTRODUCTION

Hiroshi Honda

Recent moves toward internationalization of Japanese society have led to opportunities for tenure-track positions for foreign nationals at Japanese national universities. The first employment case study follows the endeavor of Associate Professor Giuseppe Pezzotti through his moves from Osaka University to other national universities—Tohoku University, Toyohashi University of Technology, and Kyoto Institute of Technology (KIT)—and an account of how he finally came to acquire a tenured position at KIT. Other Japanese national universities, such as Hokkaido University, the University of Tokyo, Kyoto University, and Kyushu University, are currently offering tenured positions, typically for those married to a Japanese spouse. However, a majority of young Japanese scholars also have to endure a relatively long apprenticeship with their professors often of 15 years or longer until they become associate professors and are treated as junior colleagues of professors. There are exceptional cases, however, for candidates who were internationally recognized at a relatively young age through study and research abroad.

Currently, a majority of the faculty positions offered for foreign nationals at national universities are for predetermined terms and for fellowship positions funded by the ministries and their affiliate organizations in Japan, which have been introduced elsewhere in this book. Japanese private universities have been offering tenured positions for a long time, and it appears that those universities are steadily and gradually increasing their opportunities for tenure positions for foreign nationals.

The second employment case study is then-Lecturer Michael W. Barnett's account of working at Kyoto University about a decade ago. In contrast to his time there, Kyoto University has now been moving toward a graduate school–oriented structure with a significant emphasis on research, especially since its centennial in 1997, when new, additional buildings and facilities were built with governmental and other funding. Its American-style football team is one of the most popular college teams in Japan, and it celebrated its first half century in 1997, inviting the Harvard football team to compete. The team also invited major football teams of other American academies and universities, and redistributed the American players and Japanese players from both Kyoto and other Japanese universities to form new teams for the games that would be academies and universities, and redistributed the American players and Japanese players from both Kyoto and other Japanese universities to form new teams for the games that would be more balanced in strength.

Some eminent emeritus professors at Kyoto University say that the university initially followed the German pattern during its early years. The university enjoys an outstanding reputation in higher academic circles around the world in such fields as fundamental physics; however, the work atmosphere and environment are by and large still what western visitors are accustomed to, as is the case with other national universities, according to a Japanese professor there. If you can communicate in some depth in Japanese during your years there, your exposure to the culture and people at a Japanese university will become much more meaningful.

ODYSSEY TOWARD A TENURED POSITION IN A JAPANESE UNIVERSITY
Giuseppe Pezzotti

What non-Japanese generally ask me is *why* I have lived in Japan for such a long time. What Japanese generally ask me is *when* I will go back to my country. Although the only and very simple answers are that I really like Japan and that I decided to stay here permanently, it is always difficult to answer these questions. Sometimes, I just make a joke: "I was so scared to take the airplane for such a long time when I came to Japan that I don't want to take a flight back." I actually make an effort to evade the questions, trying to escape with a smile. However, it is noteworthy how both Japanese and foreigners feel that my choice of staying on "the Island" (to use an expression dear to English people) is definitely not the obvious one (some may even feel it odd). Although we have entered a new century, from a general point of view, it really looks as if Japan and foreign countries are still very far apart. Nevertheless, I will try to show that, despite the large geographical distance between Japan and its foreign partners, symptoms of imminent deep change are showing for the foreign community living in Japan and in Japanese habits and culture as well.

When I first came to Japan (12 years ago), I was certainly a typical foreigner, almost completely unaware of the Japanese culture and way of thinking and, more important, unable to speak the language. To give an idea of the mismatch between me and my surroundings, it would be sufficient to tell that I tried to put sugar into green tea. The amount of food the Japanese consume daily was so little compared with Italy, my home country, that I often astonished waiters by asking for a double portion. Although at that time I was not even considering staying permanently in Japan, I quickly (and forcedly) realized the importance of learning the Japanese language. When I became a doctoral student at Osaka University, somebody told me that research topics could certainly be discussed in English. However, it did not take much time for me to realize that my host professor at Osaka University not only did not understand what I wanted to say in English, but also that he systematically answered "yes" whatever the question was, making things even more complicated. On the other hand, despite all my efforts, my spoken Japanese remained terrible for a relatively long time. Incidentally, I should say that I never liked the Japanese-language schools or, to be more precise, the teaching method that such schools usually adopt. Since childhood, we Europeans have learned things by causal nexus (or Cartesian logic, if you prefer), hardly by a mnemonic system. Thus, teachers unable to give logical form to what they teach, whatever the subject, have never been my favorite. After being a very bad student (for a very short while indeed) in a Japanese-language school, I quit and decided to make it by myself. As a first step, I prepared several tapes in which I pronounced about 5000 words in the phonetics of my own mother tongue, followed by the Japanese pronunciation, picking them up almost randomly from a vocabulary. Wearing a headphone day and night, even while asleep, I listened to the tape in the autoreverse mode, until I could memorize a minimum vocabulary necessary for an elementary conversation. To be honest, this initial step was the most difficult in learning the language; since then, the needs of daily life and work have provided the rest.

Sometimes now, while teaching my classes on fracture and deformation mechanics (in Japanese) to my undergraduate students at Kyoto Institute of Technology, I happen to remember the time I could not speak Japanese and how strongly I wanted to learn this language. Recently, I have also happened to teach a brief summer course at Rome University (the university from which I received my bachelor's degree). I was astonished and, at the same time, worried to find myself having partly forgotten my mother tongue.

Besides the language ability, however, many things have certainly changed in my life since I came to Japan. I cannot judge whether I am better now than I was 12 years ago. However, what I know is that I will hardly be the same as I was at the time I arrived here. Life in Japan at first reduced me to the ground state, unable to tend to the most elementary needs, thus giving me a chance to develop a different, perhaps stronger, self. People change, foreigners change, and sometimes the changes are relatively quick and irreversible. This, I hope, the Japanese people will realize sometime in the twenty-first century.

I am convinced that in the Japan of 12 years ago, people had only somewhat vague ideas about Europe. For example, I remember a young lady from Osaka asking me whether the mother tongue of the Italian people was English or German. Although she was obviously not a particularly smart example of the young Japanese generation, it was true that, unlike perhaps America, Europe was very far away in the mind of the common people and was still not fully identified as a continent for a large part of the Japanese population. At present, pushed by historical and political events, Japan has started to learn more about foreign countries and to realize the importance of building up an international reputation. We have to acknowledge that, in the last decade, many things have changed in Japan. More information is available about foreign countries, and traveling abroad (leaving "the Island") is easier. I say this not only because the number of McDonald's has increased or because spaghetti cooked al dente has now become very popular. Undoubtedly, many more Japanese people are experiencing living abroad and are learning to appreciate both the merits and faults of their own country; at least they are becoming capable of discussing the faults if not yet solving them. For example, fine Japanese soccer players such as Nakata and Nanami have gone to Italy to play, and Nomo, Yoshii, and several other famous players are appreciated in American major league baseball, proving that the new generation of Japanese people can also be strong outside the

Japanese system and can work side by side with foreigners, according to common rules. All these people are important examples because their presence triggers the interest of Japanese people in the world outside. This was just unthinkable 12 years ago when I came to Japan, perhaps even more unimaginable than the present economic crisis. In brief, I have no doubts that Japan is now a much more internationalized country than it was 12 years ago.

There is another reason why Japan is becoming more internationally oriented. In several government universities now, even foreign nationals may hold tenured positions as teaching staff. (Note that the term teaching staff does not refer only to English-language teachers, whose presence is so massive in Japan as to amount to a stereotype.) Very recently, too, the government university where I work, Kyoto Institute of Technology, has made the employment rules equal for foreigners and nationals by making faculty tenure tracks available also to foreign nationals. After almost a century of medieval behavior, an old, discriminative rule has been abolished (both the old and the modified versions of the rule from the University Constitution are reproduced in English translation at the end of Chapter 10), and I became the first foreigner to hold a tenured position in the history of Kyoto Institute of Technology. Although I am certainly not the first foreigner in Japan to become a permanent government employee, a great deal of responsibility accompanies this honor, besides the gratitude to the enlightened people who made this major change possible. The importance of this change was underlined by an official visit of the first councillor of the European Union and of the scientific attaché of the Italian embassy to congratulate the university president and the faculty dean on our university's innovative stance.

So far I have recounted the official facts. It is also interesting to look at some unofficial details of the history of this change as I have personally experienced it. Undoubtedly, I had to deal with a large variety of leading Japanese people, some of them stubborn and old-fashioned (for example, some of the old directorial staff of our university, now fortunately replaced), some others very brilliant and open-minded. Hoping that my experience can be helpful to others, I feel it is worth making the unofficial facts publicly available.

Over the years, my academic career in Japan has seen plenty of ups and downs. For example, after I had received "hearty" congratulations for my doctor's degree (achieved in three years) at Osaka University and was thereupon employed as a postdoctoral fellow for three more years, my host professor told me that I had to leave, because "foreigners *cannot* get a staff position in a Japanese government university." Very naturally, he forgot the effort I made through those long years working as his English secretary and ghostwriter. Of course, he had one assistant position available, but, despite his socially high-ranked position at the famous Osaka University, he could never free his mind of old prejudices and fear of blame for being the one to employ a foreigner. In other words, just six years ago, it was considered almost *impossible,* according to Japanese common sense, that a foreigner (more particularly, a "westerner") could hold a regular position in a government university. It is possible that some foreigners elsewhere in Japan were luckier than I; however, such cases should be considered exceptions, not a trend toward change in the way of thinking in Japan at that time. Thus, I had to leave Osaka and could get only a one-year contract in a comparable position (guest researcher) at Tohoku University in Sendai, which is farther northeast than Osaka but still on Honshu island. Sendai was a new place for a new start as a foreigner, who *must* be a "guest," no matter how many years he had already spent in Japan. In addition, I ran into problems with my visa as, given the low salary of a guest researcher, I could not apply for a work visa anymore and I had to request a visa for cultural activities. This was an evident step backward in the eyes of the formal and narrow-minded employees of the Immigration Office. Of course, they had in mind the reliable parameters of Japanese society (applicable to Japanese employees only), according to which what is acquired can by no means be lost. Given that assumption, it was clear that I must be a very "bad guy" to have been treated in such a terrible way. Nevertheless, I realized at that time how Japanese people exhibit very polite and stereotyped behavior with foreign guests. This behavior may make you happy at first, but it definitely hurts and irritates you after many years of residence in Japan.

Perhaps I was not in the right mood or the right preconditions did not exist for a successful career at Tohoku University, but I was certainly no better treated in Sendai than I was at Osaka University. I was just squeezed like a lemon to produce good scientific papers, needed to cover for the low scientific level of a university staff chosen more out of natural "right" than for actual merit. Sometimes I wrote papers whose official authors I was not even allowed to know; other times, my experimental data were published by unknown people, without even being acknowledged. To my eyes, Tohoku University was not better than Osaka University, no matter how long its history or how great its prestige: it was merely an old place for old-minded people.

From the foregoing, you may think that my scientific career at that time was not very brilliant or, at least, not successful. However, given the very good facilities available in the big Japanese government universities and despite the uneven treatment to which I was subjected—even with such a handicap—I scored more than twice the number of publications in good scientific journals of the average Japanese professor.

Human relationships rather than an actually fair opportunity brought me an assistant professorship in a relatively new government university, Toyohashi University of Technology, in the prefecture of Aichi. If you carefully read the discriminative rule about foreign employees, you will see that for the lowest rank in an academic career (assistant professor), the university constitution does not foresee any discrimination in the job contract. Thus, although highly undervalued in employment level, I was very proud of having finally, at the age of 33, obtained a permanent job. About two years later, the associate professorship at Kyoto Institute of Technology looked like a very natural consequence of nine years of hard work.

A happy ending to my story, I thought. However, things were still far from a happy conclusion. In fact, given the discriminative rule in effect at that time at Kyoto Institute of Technology, being promoted to a higher position meant automatically losing my permanent contract in exchange for a renewable three-year contract. Japan is a country where everything is based on long-term plans and guarantees; there are very few things you can do to improve your social condition, if you don't have a permanent job. I thought that the injustice I had suffered should be so evident that it would win understanding from anybody, but this was not true. In Japanese society nobody seemed to care about what is right or wrong: foreigners come to Japan to leave sometime later; who cares about their rights or their future? Given the clearly unfair circumstances, for the first time I decided to play hardball.

The first time I asked the dean of the faculty of Kyoto Institute of Technology to consider the possibility of equating the job contracts of foreigners with those of Japanese employees was at the beginning of January 1997, during the university's new year party. He diplomatically asked me for some time to think about it, and, looking at his poker face, I even thought he was positively oriented toward the change. Almost concurrently, the sponsoring professor who wanted me at Kyoto Institute of Technology made a similar attempt with the university president, asking for the rule change. The president, too, was very diplomatic, but he said that he would consider such a change only if a Nobel Prize–class foreign scientist were the object of the request. No further comment about the profoundly discriminative meaning of the president's assertion is necessary. After about six months of patient waiting, I tried several times to contact the dean of the faculty to get the results of his (lengthy) considerations.

However, he was always very "busy" and could never find time for me. At the limit of my patience, after many trials, I forced him to meet me by suddenly showing up at his room. Unable to avoid me, he had to say something. What he said was very simple: "According to the general directives of the Japanese government, the discriminative rule for job contracts of nonnationals has a profound meaning, which cannot be understood at your level. Thus, the rule is still meaningful and does not need to be changed." I had no chance to reply, but I wrote down his precise words and, back in my room, I right away called the scientific attaché of the Italian embassy in Tokyo. I just asked him to officially verify Monbusho whether the simple explanation given by the dean of Kyoto Institute of Technology was true or not. My request turned the course of recent academic history.

The Italian Embassy and the European Union delegation played a very important role in the positive solution of my problem. I suppose that their thorough support was based on respect for human rights, regardless of the particular circumstances. However, we never needed to say openly between us exactly why they came to my aid. It was just obvious, because Europe went through the French Revolution and knows well how the mere interests of individuals, whatever the power they have, is negligible in comparison with human rights. On the other hand, to the older Japanese thinking of the dean of the faculty, I was just a traitor, and thus unworthy of esteem. The dean never answered two consecutive letters (the original in English and then its translation in Japanese sent to him later) from the European Union delegation, asking for clarifications about my position in the university. Then, the European Union delegation directly approached the Japanese Ministry of Education. The official answer of the ministry to the European Union was disarmingly simple: there is *no directive* from the Japanese government to discriminate between nationals and foreigners in job contracts; any government university is thus on its own in deciding on the issue for its convenience. At that time, we were ready to do anything legal to solve the problem; perhaps the most suitable way was to disclose the mere facts to the media. Probably afraid of our determination, which could cause a small scandal and create more problems, Monbusho informally contacted the dean of Kyoto Institute of Technology.

The dean's first reaction was just anger. He suddenly ordered an official meeting in his room at which a professor member of the academic senate, the department head, another professor from our department, and I were present, together with representative members of the administration office. The dean directly pointed out my bad reputation as a traitor who leaked internal facts to enemies. He added that he himself, as a public person, had a reputation to defend and that this could by no means be jeopardized by a single employee like me. He tried to find agreement among the other professors present, who were actually silent and pretty unreactive to his panegyric. I, as the only one he directly addressed, was the first who had to give some kind of answer. I simply answered that I had indeed considered the dean's reputation just as I considered how the issue was affecting my career and, ultimately, my social rights. I was actually surprised to discover that, when the dean tried to force out an answer from the other professors present at the meeting, they stepped toward my side, leaving him alone with his feverish arguments. The meeting was just a disaster for the dean; against all odds and unexpectedly (like all the good things that have been happening in Japan), I was the winner of that meeting. Of course, this could come about only because of the many nice and honest Japanese people I had around me. I learned later that the dean had tried again to press the department head to fire me, but, of course, unsuccessfully.

After some months, the dean's behavior changed considerably. Perhaps the pressure from the European Union delegation was still very insistent. Nevertheless, in December 1998 he finally changed the "wrong" rule to the shape you can now see, and, as of April 1, 1999, I became the first foreign permanent professor at Kyoto Institute of Technology. In the final vote of the professors' council, of 60 voting professors in our faculty, 59 voted favorably for my permanent contract and only one left a blank ballot. A happy ending to the story, if you will.

I don't know whether my episode had a direct influence on the dean's career. However, after these events, the dean failed his candidature to become president, failed to win renewal as dean, and even failed his candidature to become a member of the academic senate. He was put back in his professorial position—one of his old positions, at any rate—because nobody can fire a government employee in Japan. The president, too, could not get renewal of his commission.

These are the mere facts. You may think that this is not a story to air publicly, but at least it reveals what goes on behind the beautiful facade. Certainly, it belongs to the old century. I decided to make it public because, perhaps, somewhere in Japan there are others now suffering the same uneven treatment. Disclosing the mere facts may help. But it is astonishing how the Japanese Monbusho can leave such a delicate matter in the hands of local, culturally obsolete people, without showing a clear path to eliminating at least the shame of such discriminative and fully irrational behavior. Presently, at least half of Japanese government universities still actively maintain the discriminative rule for foreign employees. If Japan still considers itself to be in a vulnerable position in Asia and still mainly fears an "invasion" by neighboring countries, then there should be a different (and smarter) way to solve this problem, at least in academia. For example, Italy likewise is not in a completely secure situation in the Mediterranean area. However, no such discriminative rule is employed. The "principle of reciprocity" holds: no foreign professors from a country will be employed in Italian academia if that country may not in turn employ Italian professors—a fair treatment, adaptable case by case, avoiding any cultural racism but still invoking cultural progress. I sincerely believe that Japan possesses a wonderful culture, but nevertheless it very much needs foreigners and foreign cultures too. Here in Japan we have not had any "equal employment opportunity" to offer or clear rules for protecting minorities. We need brave hearts, not mercenaries.

I wish to close this chapter on a positive note, maybe too optimistic, but at least due, in keeping with the happy end of my story. Despite slow and somewhat erratic progress, recent changes in Japanese academia have systematically been toward better and more reasonable choices. There will always be somebody who cares. Recently, at night I have been having a recurrent dream: the twenty-first century will be the beginning of a Japanese Renaissance.

OBSERVATIONS ON WORKING AT KYOTO UNIVERSITY _____
Michael W. Barnett

Introduction: The Setting at Kyoto University

Kyoto Imperial University was founded in 1897, the second university to be established in Japan. The College of Science and Engineering was opened in September of the same year and was separated into two colleges in July 1914. After the Second World War the university was renamed Kyoto University. In 1949 it was reorganized into a four-year university, and in 1953 a graduate

---

*Japan is changing laws for equal opportunities, as can be seen in that for both sexes. (See Introduction of Part 7 of Business Version). Those for foreign nationals will come.

school was founded. As of 1991 there were 158 chairs (about 550 faculty members) in the faculty of engineering, which was organized into 23 undergraduate and 25 graduate departments, and 4 research laboratories. There were 5930 undergraduate and graduate students enrolled in engineering, of whom fewer than 3 percent were women (this percentage was typical of Japanese engineering departments). The total student population averaged around 17,000.

At its founding, Kyoto University was a state-oriented school, its main purpose being to meet the specific needs foreseen by the nation's leaders. Over time, the focus of the university has changed to address more completely the needs of society as a whole, though it has maintained a slight antistate bias in contrast to the University of Tokyo, which, since its founding, has always produced many government bureaucrats. Kyoto University has had a colorful past, having been a center for the Marxist movement before World War II and the Red Army after the war and home to more Nobel Prize winners (four) than any other university in Japan. Kyoto University is regarded in Japan as an excellent school for engineering and the sciences. In order of prestige among national government schools, Kyoto University is, arguably, number two behind the University of Tokyo.

This chapter summarizes observations I made while working at Kyoto University as both postdoctoral researcher and lecturer in the Department of Environmental and Sanitary Engineering. Perhaps it is obvious that a good knowledge of Japanese society goes a long way toward the understanding of Japanese institutions such as a university. But, because of the greater degree of homogeneity in Japan, there is more congruency among institutions than in American society. Thus, it is a little easier to generalize. Below the surface some things in Japan don't seem to change much over time. Still, when making judgments it is important to be cautious, since the situation is dynamic, especially now that Japan is finding it necessary to take a leading role in international affairs. Economically and otherwise, the international community has been rather accommodating, but now Japan needs to reciprocate. Ensuring a successful transition to a more open society is one of the challenges facing Japan today.

Form and Substance

On the surface, Japanese universities can appear to be much like western universities. The university's mission in society is clearly defined, the administrative organization chart shows several faculties and institutes overseeing various university activities, classes appear to be western-style, and research laboratories are much like those in the United States. Yet, close inspection reveals many differences. The university structure is rigid, there is little movement of persons or ideas horizontally, the promotion system is inflexible, the quality of lectures varies quite a bit, and the behavior of both teachers and students in and out of the classroom is completely different. In fact, the contrast is so great that it is tempting to conclude that many activities are rituals performed for no other reason than that it is written somewhere that it must be done this way. Outward appearance is not a good indicator of what goes on inside, and there are significant problems; however, in many ways the system works. The education system, of which the university is only a part, does develop competent engineers and scientists, but the manner in which this is achieved differs considerably from the way it is done in the west.

It is well known that since the war the interests of the economic producers have been highest on this nation's priority list. It has also been noted that Japanese industries in their recruiting practices place a lower premium on specialized skills than do western industries, and these practices have in turn affected the nature of university education in Japan.[1] Emphasis on the economy, combined with the need to supply industry with persons having "latent ability," has cast universities, especially engineering departments, in a specific industry support role and left less room for purely academic pursuits. American universities stress specialized, individual professional development and creative, scholarly research, with companies largely being forced to accept what they get. In contrast, Japanese universities emphasize cooperation, coordinated development of needed skills and technology and are sensitive to the demands of business and industry. Industry, for its part, has shown a willingness to shoulder much of the responsibility for training and thus is an important part of the education system in Japan. This appears to be one of the strengths of the system.

University education stands in sharp contrast to primary and secondary schooling in Japan. The characteristics of basic education in Japan have been documented, as have those of Japanese universities.[2] One problem in the university concerns the students. Some, even at the master's level, devote more time to extracurricular activities than would normally be permitted at an American

---

[1] Robert C. Christopher, *The Japanese Mind: The Goliath Explained*, Fawcett Columbine, New York, 1984 (also Ballantine, New York, 1984), p. 93.

university. Young Japanese know that acceptance to the right university, by passing a series of rigorous examinations, will both ensure them a good job and allow them four or more years of free time to devote to club activities as well as, for the first time since early childhood, enjoying life. This often becomes a consideration when deciding upon continuing education to the graduate level. A large fraction of students take the entrance exams three or four times (exams are given once a year by each university), since many students apply for admission to more than one university. There is little or no movement laterally between departments, and students who pass the examinations must decide early which career to pursue. At Kyoto University, freshmen are accepted into a specific faculty and, in the case of engineering, a specific department. The system is rigid, but on the other hand, it can be argued that the average Japanese student has a solid foundation in basic learning skills, having struggled for many years in elementary and secondary school. Skills are uniform among all groups of students majoring in engineering, and the educational level of entering freshmen is probably higher than in the United States. There is also uniformity in the sense that everyone seems to know the same things. Rote learning in primary school is heavily reinforced, for example, as needed to master the complex writing system. Individualistic tendencies, including certain creative activities, are generally frowned upon in the university as in Japanese society as a whole. Conformity to the group is considered best.

The manner in which lectures are conducted in a Japanese university often shocks outside observers. In the social hierarchy a professor is held in high regard and it is impolite to question his or her authority. Students are nonaggressive and reserved in their behavior. Few, if any, questions are asked in the classroom. The teacher prepares notes for distribution and delivers lectures uninterrupted. It is difficult to get Japanese professors to lecture to groups of foreign students, since Japanese professors are uncomfortable with the confrontational character of western-style teaching, where student and teacher may challenge one another. There is much talk about "challenging the student to learn," but here little can be done without some cooperation from the students. There is a kind of nongrading grading system in which only the worst students are given failing grades and essentially all undergraduate students finish in four years; master course students, in two years.

---

2 Edwin O. Reischauer, *Japan: The Story of a Nation*, 4th ed., McGraw-Hill, New York, 1989, pp. 186–202.

Administratively, the department is a blend of western-style rules and procedures and Japanese-brand bureaucracy and decision making. Weekly meetings are held to discuss department business, but decision making calls for many behind-the-scenes discussions with affected groups. Important decisions are effectively made before being presented to others at the faculty meeting. Junior members defer judgment to full professors, and a professor's judgment is never questioned. In a very real sense the department is like a family. As is typical in Japan, 80 to 90 or more percent of the faculty members are previous graduates of the department in which they currently work. A hierarchy of prestige is maintained thus; for example, much attention is given to the ordering of names on official documents. Many activities are tightly controlled by Monbusho. Changes in personnel, such as promotions to professor, or curricula, such as addition of a new class, become political issues and may take months or years. The degree of control exerted is not always in proportion to the amount of funding provided by Monbusho. A professor (the person responsible for obtaining and distributing funds through his or her chair) who chooses to obtain financial support from industry or business may, if successful, fund most of the chair's research activities from these sources. Funds for overhead, such as building upkeep or remodeling, and other indirect costs are insufficient, and university salaries as a whole are low compared with those in American universities. Outside consulting (by employees of government institutions) is not permitted, but with the blessing of one's professor, honoraria can be collected from companies for giving lectures or technical advice.

The quality of a given chair is highly dependent on the quality of the professor in the chair, since this person, like the head of a family, is the primary funding source and decision maker (sometimes including decisions regarding employment of graduating students through the chair). The system permits qualified individuals to move to the top but also drags with it some poorly motivated others. The essential policies of those wanting to advance in the system are to do what you are told, never question your superiors, and persevere. Eventually, the system will reward those who are patient. Good professors know this and grant promising individuals the funds and freedom to do their work. Bad professors take advantage of the situation by dictating to their staff and students, stifling creativity, or just doing nothing. The system permits all these behaviors. It is possible for the complexion of a department to change considerably with personnel changes or the mandatory retirement of a professor (at age 63 for Kyoto University). Power can become centralized, and because professors are held in such high regard, one professor can have a great influence on work in an entire discipline. The professor is always listed as coauthor on publications, regardless of actual contribution.

Too often appearances are weighted more heavily than substance, but the overall goal of excellence in education and research is not forgotten, though it is sometimes set aside. The working environment is generally good, though the facilities are somewhat old and cramped, and as in American universities there is freedom in setting one's schedule. If a niche can be found in the group, effective teaching and research can be accomplished. It is wise to invest some time finding this niche before deciding to jump into the system.

At the Altar of Internationalization

The president's foreword in the 1988–89 Kyoto University Bulletin concerns international academic relations. The president was well known for his activities in this area and was a great supporter of exchange with foreign countries. His policies had been successful in creating a good environment for international cooperative exchange at Kyoto University.

It is possible to hear something about "internationalization" several times during a normal day in Japan. It is a word that has been on the lips of every Japanese for several years, having been placed foremost in the public consciousness by government leaders who saw the need for greater understanding of other countries and cultures. It can be witnessed in many forms, such as the increased availability and diversity of imported goods (at a price) and the adoption of foreign ideas. Kyoto and other large cities in Japan are very international places—more than many American cities, for example, those in the Midwest.

There are two faces to internationalization, one good, one bad. Internationalization has been successful in Japan, though there is still resistance to change. As mentioned previously, the university is quite internationalized and many foreign students, researchers, and visiting scholars can be found on campus. Prominent scholars often visit the university to make presentations. Though the university environment may not be typical of Japan as a whole, there are genuine feelings toward and desire to learn from foreign visitors. It is possible to have candid discussions with colleagues on most topics, and the majority understand the difficulties of adapting to a new culture. Japanese are good hosts and will typically go to great lengths to help you if you have problems.

The dark side of internationalization hides a significant problem in Japan, namely, strong feelings of separateness and uniqueness. Of course, it is good to feel that one's culture is unique; however, in Japan these feelings are intense, as though the uniqueness itself were unique. Japan, willingly, feels separate from the rest of the world; thus, the world feels separate from Japan and reacts with distrust.[3] Japanese are being pulled in different directions, finding it necessary to be both open, that is, to be internationalized, and closed, that is, to maintain their Japanese-ness. The tension has resulted in a degree of neurosis in Japan. At the university, this is felt in several ways. Students, researchers, visiting professors, and others are invited without adequate consideration for work they might conduct or how they might fit into the group, papers for many "international" symposia in Japan are solicited but never fully appreciated or even read, and qualified technical professionals find themselves occupying slots as token foreigners and English teachers. One sometimes feels as though one were being sacrificed in the name of internationalization. Many problems are related to language and cultural differences, and the foreign visitor can lessen such problems through his or her own study. Still, it is necessary to be wary of those who are not truly being open or sensitive to your professional needs. Fortunately, many in the university are aware of the problem. Since they must work within a system that is not always accommodating, it is important to be patient, but most problems can be worked out (in a Japanese manner) to the satisfaction of both parties.

Research and Study in Japan: Advice to Visiting Professionals, Researchers, and Students

The best advice to give to individuals interested in coming to Japan for work or study is to plan ahead. It is difficult to "parachute" into Japan and conduct meaningful work without a certain degree of luck. The preferred route is a preplanned one which is the result of both sides' having been familiarized with each other's work, perhaps as a result of a previous association, and having discussed specific goals and objectives of the work to be done. I list here a few specific points with the hope that the reader may find them useful in planning work or study in Japan:

[3] Edwin O. Reischauer Center for East Asian Studies, Paul H. Nitze School of Advanced International Studies, Johns Hopkins University, *The United States and Japan in 1990: A New World Environment, New Questions*, Japan Times, Tokyo, 1990, p. 89.

▶ Be patient. Things take time to develop. Make an effort to become part of the work group, and have work you can do on your own so that, if you must wait, your time won't be wasted. It may take some time to coordinate your activities with those of your laboratory, and you may begin to feel frustrated. Unfortunately, some professors do not make an effort to assimilate foreign visitors or facilitate their study or research. Avoid these professors. Some visitors have been known to give up entirely and devote their energies to other activities. If you find it impossible to do work in your field, spend your time studying the language and culture.

▶ To the degree possible, try to learn the Japanese language, preferably by completing course work before coming to Japan. Effective language teaching methods are generally not in widespread use in Japan, and language schools are expensive. Surprisingly, some experienced students of Japanese think it is difficult to learn Japanese in Japan. While in Japan, budget some time for language study and remember that a high degree of cultural sensitivity is vital to effective communication. It is essential to develop, in any language, a good rapport with your work group. Go to parties and social functions, resist the temptation to separate from the group, and maintain a positive attitude.

▶ Japan was expensive and pay scales were generally lower in the early 1990s (The situation is better as of 2000 for academia). Also, housing is a serious problem. Your standard of living may be lower than in the United States. Be ready to make some sacrifices. If you choose a Japanese style of living, including the food you eat, it will be easier.

▶ It is a practical reality that, in and out of the university, full professors are powerful people; thus, you should make an effort to cooperate fully with them. Try to understand and work within the system. Don't make unreasonable demands.

Summary

The experience of research or study at Japanese universities can be either good or bad depending on the adaptability of the foreign visitor and that of the laboratory in which he or she decides to conduct work. Universities are only a part of the educational system in Japan, which includes primary and secondary schools as well as business and industry. There is opportunity at Japanese universities for effective research and study, but foreign visitors should take care in planning their stay. The Japanese way of doing things may seem strange at first, but it does have many advantages. Japanese are culturally conditioned for hard work, sacrifice, and perseverance and prefer cooperation over conflict. Yet, the system is rigid, creative tendencies are suppressed, and society is still not completely open to outsiders. Internationalization is having a positive effect, and Japan has certainly shown itself to be adaptable. There is much promise in cooperative work with Japan; it is our responsibility as guests to do our best to understand the system in order to facilitate cooperative efforts and strengthen our ties with this important member of the global community.

I would like to thank my friends and colleagues who took the time to comment on this chapter. Dr. Masafumi Goto of Kajima Corporation–Marine Biotechnology Institute Co., Ltd., Dr. Bernard B. Siman of Jardine Fleming Securities, Ltd., and Mr. Paul Driscoll of the Kyoto University Research Center for Biomedical Engineering were especially helpful. Thanks to Professor Masakatsu Hiraoka and his staff in the Department of Environmental and Sanitary Engineering at Kyoto University. Finally, I would like to thank Dr. Charles W. Knisely for introducing me to Professor Wataru Nakayama of the ASME Japan Chapter, and Dr. Hiroshi Honda for giving me the opportunity to make a contribution to this book.

References

James Fallows, *More Like US,* Houghton-Mifflin, Boston, 1990.

Ivan P. Hall, "Organizational Paralysis: The Case of Todai," in Ezra F. Vogel, (ed.), *Modern Japanese Organization and Decision Making,* University of California Press, Berkeley, 1975.

Karel van Wolferen, *The Enigma of Japanese Power,* Knopf, New York, 1989 (also Random House/Vintage, New York, 1989).

23

Establishing Firms in Japan

Stephen A. Hann and Hiroshi Honda

GENERAL NOTES

Hiroshi Honda

Economic slumps in Japan from 1997 to 1999 prompted many employees of Japanese companies to leave their places of employment and start their own new ventures. Governmental organizations such as MITI encourage and support once-loyal salaried persons starting their own businesses, since they expect that this trend will help revitalize Japan's economy, especially in challenging and promising new fields. It is now relatively easy for anyone to establish a new company, and it is widely recommended that new entrepreneurs establish a limited company *(yuugen gaisha)* as a starting point. Joint-stock companies *(kabushiki gaisha)* can also be established, but they take more effort and require complicated procedures to establish them. Table 23.1 shows comparative features of the two kinds of companies. Some people who have already established small joint-stock companies later convert them to limited companies so that they will be prepared in case they need to scale down business operations to cut costs, and/or to make them mobile in management.

If your Japanese-language competency falls short of that of educated Japanese people, it would be wise to seek professional help for drafting the necessary documents to establish a company. Among the reasons for getting help are that the wording of the documents is legalistic, and some of the traditional wording might appear unnatural to you from a scientific and technological viewpoint. The additional cost of using professional help could range from 500,000 to over

Table 23.1 Comparative Features of Yuugen Gaisha and Kabushiki Gaisha

Item	Yuugen Gaisha (Limited Company)	Kabushiki Gaisha (Joint-Stock Company)
Minimum capital	3 million yen	10 million yen
Unit share	50,000 yen	50,000 yen
Registration fee, etc. (Apr. 1998)	Minimum 150,000 yen	Minimum 240,000 yen
Maximum number of investors	50	Unlimited
Investors' responsibility	Own investment amount	Own investment amount
Minimum number of members of board of directors	One	Three
Appointment of auditor	Unnecessary	Minimum one
Merits	Simplified procedure for company establishment	Much investment can be collected

1 million yen (in some cases such as those involving significant interpretation and translation work), depending on the individual situation. Finally, acceptance of your registration is based on past registration records; you cannot register a new company under a name that infringes on the name of an existing company registered at the same regional legal affairs bureau.

The procedure for establishing a limited company is as follows:

1. Check with the Regional Legal Affairs Bureau where you are planning to register your company to find out whether the proposed name of your company is already registered by any third party.
2. If it is not, construct the articles of your company in the Japanese language. You need to appoint a minimum of one member of the board of directors, and an auditor (for limited companies, not mandatory), before you construct the articles.
3. The articles must be certified by a notary public with an office in the same region.
4. Deposit the minimum capital in a bank (or other financial institution) that has agreed to receive the capital for your company. (Prior contact with the bank is recommended.)

5. Construct the registration documents to be submitted to the Regional Legal Affairs Bureau.

6. Apply for registration of your company with a submission of your registration documents to the Regional Legal Affairs Bureau. (The submission must be completed within two weeks after deposit of the capital with your bank or other financial institution.)

7. After registration of your company, the predetermined document forms must be filled out and submitted to relevant government offices such as the Regional Tax Office, the City Office, the Public Employment Security Office and the Labor Standard Bureau (unnecessary if you hire no employees), the Social Insurance Office, and the bank.

The procedure for establishing a joint-stock company requires further steps, such as an offering to stockholders and sending invitation letters of general meetings to founders and stockholders. As implied before, it is possible to convert a limited company to a joint-stock company and vice versa. Therefore it is best to establish a limited company first if you are starting a small business. If your company is registered in countries outside of Japan, its registration in Japan will be relatively easy.

It may take a few years after starting your business before you are blessed with regular customers, unless you have specific products or services for which there is a market demand, such as attractive computer software developed in the west, especially Internet-related.

The following case study is the story of an American entrepreneur, Stephen A. Hann, who has had a successful software business in Japan.

AN ENGINEERING CONSULTING FIRM
Stephen A. Hann (Updated by Hiroshi Honda)

Most discussions about foreign business opportunities in Japan fall into two categories. Individuals who have started businesses in Japan that have succeeded maintain that Japan is the most open and competitive market in the world. However, those who have started businesses in Japan that have failed complain that Japan is a market closed to all foreigners and monopolized by a handful of insiders. Since my consulting practice in Japan has continued from

1989 to date. I consider my company to belong to the first group. This section is my attempt to describe my observations in Japan as objectively as possible.

My discussion is restricted to what I have actually seen during my efforts in building a mechanical engineering consulting service in Japan. This section has three parts: my impression of the business climate in Japan for engineering consulting; a report on how and why my consulting practice was started in Japan; and a list of 10 of the problems that a business will face in setting up in Japan. It is my intention that the third section be the most useful and interesting. This is information that I did not have when I started business in Japan. Had I known these things, I would have structured my business differently and planned it a lot more carefully. Furthermore, I do not discuss the immense rewards for building a successful business in Japan. This is because the required commitment of time, money, and perseverance to achieve anything there is a subject that must be thoroughly understood by anyone with any business ambitions in Japan.

It is my opinion that simply earning a lot of money in a short period of time is not a good enough reason for moving to Japan. For one thing, I haven't seen or heard of anyone accomplishing this. What I have seen is people grinding out a living through what, over a number of years, sometimes develops into a successful business. I want everyone reading this section to understand and concentrate on the challenge of breaking into the Japanese market. The business practices that impede a new venture that is getting started also work to ensure success for those companies that do establish themselves. It should also be kept in mind that only a small percentage of businesses, even larger and well-financed ones, survive the first few years.

I highly recommend two books that will be useful to anyone who is interested in starting any kind of business in Japan. A book that directly addresses the concerns of small businesses is *Setting Up and Operating a Business in Japan,* by Helen Thian (second edition, Tuttle, Tokyo, 1989, ISBN 0-8048-1544-5). While it discusses many items more of interest to importers of consumer goods than to engineers, it is a good overview of the potential rewards and of certain problems that are involved in setting up a business in Japan. Another view can be found in *Competing in Japan: Make It Here You Can Make It Anywhere!* by P. Reed Maurer (Japan Times, Tokyo, 1989, ISBN 4-7890-0486-4). Maurer's contention is that all of the difficulties of doing business in Japan are really opportunities and that Japan is the ideal marketplace.

Opportunities for Engineering Consulting in Japan

While it is true that the Japanese distribution system is resistant to outsiders, I have seen absolutely no discrimination as a foreigner trying to sell services to Japanese firms. This is because the resistance extends to all outsiders, including the Japanese themselves. A Japanese engineer trying to start a business in Japan would face every obstacle (except visa problems) that a foreigner does. In fact, foreign engineers are probably treated a little better and have somewhat fewer cultural problems than native engineers. I say this on the basis of what I have seen while trying to build a business that does not directly compete with established companies in Japan; if I had been trying to directly import and distribute foreign cars or computers, my observations might not be the same!

There are relatively few small engineering consulting firms in Japan. The reasons for this are unrelated to the amount of work available and the need for their services. In fact, I am not sure why such a large market has not attracted many more native engineers and firms. For whatever reasons, there is a large, virtually untapped market in mechanical engineering services. There is a real shortage of engineers who are highly skilled in mechanical computer-aided engineering (MCAE) software—software using finite elements, solids modeling, and nonlinear rigid-body dynamics—and who can consult on site. Currently, American and European consultants are filling this void, but they are good only for full-time on-site contracts. They come to Japan for the length of the contract—typically a week to six months—work full-time on site, and then leave. If a company wants an on-site consultant only part-time, these foreigners can do nothing for it. A Japanese MCAE distributor may want a consultant available for infrequent but important sales calls and benchmarks, but contractors cannot travel to and from Japan in such a chaotic fashion.

There is an opportunity for foreign engineers who are willing to move to and live in Japan. They must be willing to travel within Japan and work on site. This is a tall order. Most engineering R&D is not centralized in Tokyo but spread out all over the country. On many on-site contracts I spend as much time traveling as I do working. This is not a financial problem; the client companies expect their guests to be traveling from Tokyo and are willing to build the time and costs of travel into the consulting contracts. The problem is the wear and tear on the consultant, but traveling is an important part of providing consulting services and cannot be ignored.

While there are other opportunities for foreign engineers in Japan, on-site consulting and distributor support are the areas of most interest to me. Because of the lack of competition and my ability to meet the needs of the clients, there has been no reason for me to try any other fields. The rest of this discussion will be oriented toward this business.

Establishing a Consulting Business

I first visited Japan in 1987 for a total of three months. I was working in the consulting group of an MCAE software company. Until then most of my projects had been done in the home office and my major customers were local firms. This was my first overseas trip, and it looked like an isolated contract that would lead to no other new business. However, when I arrived in Japan I had a chance to talk to some of our software customers who wanted consulting services but did not want to bring a consultant over from the United States. They wanted to deal with a local firm. The local software distributors had no interest in going into the consulting business. After returning to the United States, I tried to talk my company into sending me over to Japan to start a consulting and distributor services branch office. The company was not interested. The timing was extremely fortunate for me. I had thereafter decided to quit my job and start my own consulting business and was trying to decide how to find my first contracts. There were ethical considerations involved in contacting former clients, and on the other hand I had no idea of how to sell consulting services to complete strangers. Now, here was a market asking for just the services that I wanted to provide and my company had turned it down. It seemed like the perfect situation for starting a new business. In September 1988 I called my new prospective customers in Japan and tried to get some contracts. The client companies wanted my services but would not commit themselves to contracts until I moved to Japan. The software distributors were a little more cooperative and were willing to give me a retainer to support their software sales and act as distributors for my consulting services. I picked a sole distributor over the Christmas vacation and moved to Japan in March 1989. The whole process took about six months from my first serious inquiry to finally moving to Japan.

It was a real shock to find out how slowly things moved once I arrived. The retainer from the distributor was what initially kept me in business. By the end of 1989 I had billed only 30 days of consulting to clients. The business finally started arriving in January 1990. Since then I had been busy except for three bad months. It looked as if my business was reasonably well established and it was time to start looking for longer-term and larger contracts. I was also looking for my first new employee. To avoid visa problems and to help me with language problems I decided to hire a native Japanese engineer.

Ten Problems in Setting Up a Small Business in Japan

I have learned a great deal about running a small business in Japan. Here are some items that anyone considering setting up a small business there should look into further:

1. *Setup costs in and around 1991 were extraordinarily high in Japan and became reasonably lower in 2000.* While this is the standard advice that foreigners receive when they move to Japan, I was surprised to find out just *how* expensive in 1989. A reliable estimate was that everything, except for office space, costs twice to three times what it did in the United States. Office space in Tokyo, YoKohama and Osaka areas is available, average monthly rents are shown in Appendix 6. Renting office space requires a guarantor (a small company cannot guarantee itself) and on the order of half to two years' rent as deposit (sometimes not refunded). I work on site, at my apartment, and at my distributor's office in order to avoid leasing a separate office. I also live outside of the 23 ward of Tokyo. Anyone who has not budgeted for these costs doesn't stand a realistic chance of succeeding in business in Japan.

2. *Everything takes more time in Japan.* While this is also standard advice for newcomers, it is also accurate. I thought that I had large contracts waiting for me when I arrived in Japan. That was not the case. The work that I actually received from my previous correspondence was about 15 percent of what was promised and it took three months for it to start. Anyone who comes to Japan with less than six months' living expenses (one year would be better) is asking for failure.

3. *American registration is important.* I was surprised how relieved prospective customers were when they found out that my business was incorporated in the United States. Apparently, they were not geared to working with individuals; to do business with the Japanese, it is necessary to be representing a corporation.

4. *There will be visa problems.* The Japanese government is not geared to admitting the employees of foreign small businesses. Since I was not employed by a major multinational corporation, my application for a three-year business visa was rejected without even being considered. I must renew my working visa annually. Furthermore, the only reason I was granted a working visa is that my distributor (a subsidiary of the largest media corporation in Japan) acted as my main guarantor. A small foreign business cannot act as a guarantor for its employees. All foreign businesses have some sort of visa problems Japan.

5. *Discounting does not increase business.* There is no way to get contracts any faster than just waiting. Japanese companies expect to pay full price for services, and offering a discount will not move them to you any faster.

6. *It is impossible to sell goods or services without a distributor.* Major Japanese companies will not deal directly with a small foreign business. They want to deal with a Japanese distributor in whom they have confidence. Trying to start an engineering consulting business in Japan without a local distributor is suicidal.

7. *Choosing a distributor is the most important decision that a small business makes.* I cannot find my own contracts; only my distributor can find them. I am dependent on my distributor for guarantees for my apartment and my working visa. Any increase in my business is more a sign of how actively my distributor is looking for consulting contracts than anything that I have done. A small business that chooses the wrong distributor starts out with two strikes against it.

8. *Customers are extremely demanding with service companies.* It is amazing how many meetings I am asked to attend and reports I am asked to write, sometimes on short notice. In Japan, the MCAE software distributors perform benchmark problems free that an American or European company would treat as consulting and charge for. While we all charge a great deal for our services, we earn it with the many services that we provide.

9. *It may be difficult to hire Japanese engineers.* The engineers that are there would much rather work for a large and established Japanese firm. However, there is a chance that we can find an engineer in the current trend of tough employment condition for engineers in Japan as of 2000.

10. *Not speaking Japanese is a problem but not an insurmountable barrier.* I am working with English software with associated English manuals. Most of the engineers in Japan read and write fairly well in English and

some also speak it rather well. Not learning the language has been my greatest failing in Japan, and I recommend that others not make the same mistake.

I left Japan for Ann Arbor, Michigan in 1994, when the U.S. economy became strong and Japanese economy slowed down. However, I still continue business in Japan and other Asian countries and have intermittently visited there on business as of 1999.

In closing, I hope that anyone with an interest in starting an engineering consulting practice in Japan does not think that this section exaggerates the problems in opening a new business there. These problems are some of the reasons so many new businesses, both domestic and foreign, fail in Japan. If anyone feels confident about beating the odds, Japan is a good place to start a mechanical engineering consulting practice.

PART

7

Conclusions

24

Career Opportunities after Japan

Hiroshi Honda

INTRODUCTION

The more Japanese companies, private universities, and private- and public-sector organizations (such those introduced in previous chapters) have become involved in global operations, the more job opportunities both in Japan and the rest of the world have become available to foreign-born professionals with the experience of working in Japan. A result of Japanese internationalization is that companies, universities, and organizations of countries other than Japan have had increasing interactions with their Japanese counterparts. Therefore, foreign-born professionals who have worked in Japan now also have more job opportunities with *non*-Japanese organizations both in Japan and in the rest of the world.

Having been associated with many foreign born-professionals who are currently working and/or have worked in Japan, I can see a variety of career opportunities for these professionals after Japan. However, the opportunities for them will vary, depending on their skills, their nationality, the nature of their professions, their social environment, and their own personality. Those who are from wealthy nations such as the United States and Western Europe, where there are numerous job opportunities for international dealings and businesses, have many advantages over others who are from nations that may not have the capacity to offer decent job opportunities even to their own nationals. (For example, I was once told by an Indian science professor that, should an Indian university pay even a fraction of his salary at his American university, he would go back to India to return the benefit of what he had learned to his mother country.) Taking this fact into consideration, I would like to recount some stories that illustrate the aforementioned issues from my perspective.

SOME STORIES

If you are working at a Japanese subsidiary of a company from your home country and/or with a joint venture of a Japanese company and a company from your home country, it is important that you regularly report to your superiors about your experiences in Japan. Those who know your competency and personality will probably be able to offer you job opportunities after Japan ranging from reasonable to excellent.

If you are in a top-level management position, projects and/or work you are in charge of can lead to a future job opportunity. Carlos Ghosn, a French national, born in Brazil, and the first non-Japanese president and chief operating officer of Nissan Motor Company, found the opportunity for this position through seeking a capital alliance with Nissan. It was reported that Mr. Ghosn, having worked at a Latin American subsidiary of Renault, felt that Nissan would need him when he was dealing with Nissan as a representative of Renault. It is said that he will have an opportunity for a higher-level position at Renault, if he can successfully build up Nissan's business.

Jon Elmendorf, the author of the chapter corresponding to this one for the first edition of this book, was on a career track inside the Westinghouse-group companies. By regularly reporting to his mentors within his parent organization concerning his progress, problems, and career interest, he could sense what the prospects were for future career opportunities. He was offered three attractive positions within the Westinghouse-group companies after a successful period as president at Westinghouse Energy Systems–Kobe, Japan.

Jeff Larsen was sent to Japan for a joint business development program between Stanford Research Institute (SRI) and the Daiichi Pharmaceutical Company (DPC). He has gained overall knowledge of the pharmaceutical business and extensive human connections in Japan through his work at SRI and DPC. Upon his return to the United States, he continued consulting for the pharmaceutical industry and cofounded the Health Strategies Group based in Palo Alto, California, of which he currently serves as managing director.

David Jones was hired by ESTECH Corporation, a joint venture between Nissan Motor Company and a U.S. computer software company. He had a master's degree in mechanical engineering and decided to go to the University of Chicago to acquire his M.B.A. degree after working in Japan. It appears that extensive job opportunities are available to him for positions such as business or engineering manager or director.

Some people of my acquaintance were hired by the headquarters of Japanese companies that have a high reputation for their products and have subsidiaries and related agents throughout the world. After a few years of experience of working in Japan, some of them were offered managerial positions at those subsidiaries. These jobs are often offered to those who could adapt to the Japanese work environment, rather than on the basis of specific job skills. Managers of Japanese companies often say that adaptability is essential, specific job skills, preferred.

It is also important that your work and professional competency be well recognized by scientists and engineers in your field. Contributing scientific and technical papers to appropriate journals in your own and relevant fields and/or articles to a book like this one can lead to international recognition and subsequently to job opportunities, often offered in an unexpected manner and by unexpected persons. Professor Oliver Wright, a British solid-state physicist, had worked at CNRS in Grenoble and at the Schlumberger research center in Paris, prior to coming to a research laboratory at Nippon Steel in Japan. He married a Japanese woman, and his Japanese-language skill reached the level of passing the grade-one Japanese Language Proficiency Test established by the Japan Foundation. He left for Europe to work at CNR in Italy in 1994, but returned to Japan in 1996 as a professor of applied physics in the Faculty of Engineering of Hokkaido University. His international recognition in his field, as well as his adaptability to Japanese society, made these moves possible.

Dr. Joyce Yamamoto, a Japanese-American materials scientist, initially came to a Japanese national laboratory on a postdoctoral fellowship. At the end of the fellowship, she worked for a private company as a visiting researcher. Returning to the United States three years later, she was a staff researcher at Cornell University. Since that time she has worked for Motorola in research and development.

If a scientist possesses a solid scientific knowledge and competency in conducting research in his or her own field, he or she can work at fairly extensive places in the world.

Being a Middle Easterner and having been educated in the United Kingdom and the United States as well, Dr. Ghassem Zarbi had to make a significant effort to get into and adapt himself to the work environment at a research laboratory of a Japanese company after receiving a doctor's degree in engineering from Sophia University, Japan. Japanese companies and public- and private-sector organizations generally prefer to hire new graduates and keep them throughout their entire career. Those who enter the companies with previous job experience are often regarded as outsiders and would not be treated equally with Japanese regular employees of the same qualifications or even of lower qualifications. (I have heard of this kind of treatment, perhaps to a lesser degree, even among some of the leading nations of the west.) It can happen, however, that the salaries of the foreign-born professionals are higher than those of their Japanese counterparts, especially in the case of the westerners, for the reason that they are hired only on a temporary basis. However, those outsiders can build up job experience and expertise. After a total of about 10 years of conducting graduate studies at a Japanese university and working in Japan, Dr. Zarbi has gone to Canada with his wife and daughter to find a job. His experiences in Japan have led to his development of new products at a Canadian company. After three years of successfully working in Canada, he bought a new house and is now planning to become a Canadian citizen. Being trilingual (Persian, English, and Japanese), Dr. Zarbi may have an opportunity to serve as an engineering and business manager or director in the future.

Those who are sent to Japan by universities and governmental organizations in developing countries will likely have a fast-track career opportunity within their organization. Even if they decide to switch to positions in the private sector from their positions in the public sector, they will still have good career opportunities. This does not seem to happen too frequently in developing countries, where the status of government officials or university professors is very high and rewarding especially for those countries.

SUMMARY REMARKS

It is very important to build up your skill as an engineer and a scientist if you wish to find career path opportunities in management and/or a professorship. During the process, frequent reporting to your mentors is also important. You will never know what particular experiences among those you had in Japan could lead to job opportunities after Japan. You may find jobs through advertisements in newspapers, journals, and other media, including the Internet. Your references will play an important role in finding a job via any means. Within the scope of your professional activities and your capacity, it would be wise to constantly make efforts to broaden, deepen, and refine your job skills, interpersonal skills, human connections, professional outlook, and foresight.

Recognition is also important in this age of global competition. Writing of some of your experiences of working in Japan for journals (those of private companies of commercial nature, or of professional societies) is to be encouraged. Extensive connections with people at different levels are also important, even though you never know who will bring you opportunities until they happen. Networking within the alumni association of your alma mater is also helpful.

25

Concluding Remarks

Hiroshi Honda

Since the first edition of this book was published in 1991, companies, universities and other organizations that hire foreign-born professionals have had extensive experience dealing with them. By the same token, those who were hired have also had extensive experience of the work environment in Japan, of working with Japanese colleagues, and of meeting and collaborating with Japanese people outside their organizations. Through their experience, these organizations, especially private companies, came to find it important to hire candidates with a high adaptability to the Japanese work environment. Personnel departments and managers that were hiring foreign-born professionals often even identified job skills and communication skills in Japanese as a second priority. Of course, there are exceptions, for positions that require exceptional job skills and high-level academic and research competency, such as high-level management jobs or professorships. These organizations also came to recognize through experience the importance of matching the organization's interest and the candidate's interest. Over time, those who liked the Japanese work environment tended to come back to Japan even after having to leave the country for some time.

To get a sense of current trends in hiring practice and preferences among different kinds of organizations, the author sent out a questionnaire to selected leading Japanese companies, listed in the first section of Tokyo Stock Exchange Market, of the industry sectors, shown in Table 25.1, in August 1999.

The hiring practice varied extensively depending on industry sector, but the current trends could be summarized as follows, based on the results obtained:

1. Among the companies that replied, 87.5% of them are currently hiring foreign professionals,
2. Among the companies that replied, 68.8% and 43.8% of them plan to hire foreign professionals and foreign woman professionals, respectively. (One company stressed that they had an equal employment opportunity policy for both sexes.)
3. The numbers of foreign professionals hired by the companies that replied are found to range from 2 to approximately 100, with an average of 26.8 employees per company. The numbers in their overseas subsidiaries could

Table 25.1   Industry Sectors for Survey on Employment of Foreign Professionals

No.	Industry sectors to which the questionnaire was sent	0: Industry sectors that replied
1.	Automobile	0
2.	Bank	0
3.	Chemicals	0
4.	Communication	–
5.	Construction	0
6.	Electric Power and Gas	0
7.	Electrical Machinery	0
8.	Electronics and Computers	0
9.	Heavy Industry and Machinery	0
10.	Insurance	0
11.	Land, Marine and Air Transport	–
12.	Mining	0
13.	Nonferrous Metals	0
14.	Oil and Coal Products	0
15.	Pharmaceuticals	0
16.	Rubber Products	–
17.	Securities	–
18.	Services	–
19.	Steel	0
20.	Warehousing & Harbor Transport Services	–
21.	Wholesale	0

range from zero to more than ten thousands, depending on the scale of the overseas business operations.

4. Among the companies that replied, 12.5%, 43.8% and 31.3% of them noted that their average employment periods of the foreign professionals are two to three years, four to five years, and more than five years, respectively. (The average employment period was reported to be two to three years in 1991, according to the first edition of this book.) Some companies also commented that the employment period varies extensively, depending on the individuals.

5. Some domestic business-oriented sectors such as electric power, gas, local banking, and marine & fire insurance sectors do not appear to regularly hire foreign professionals at their domestic offices. Mining-sector companies have overseas business operations; however, they also do not appear to regularly hire foreign professionals at their domestic offices. (Exceptions can be seen in cases when governmental sectors are involved as part of ODA programs, or overseas organizations in business alliance, etc., are involved.) However, companies in these sectors will certainly hire foreign professionals at their overseas subsidiaries and/or affiliated companies. This trend is especially marked in financial-sector companies, which will need local professionals to handle mergers and acquisitions, loan, securities, bonds, etc.

6. The leading established Japanese companies tend to set the same hiring standard for all candidates of all nationalities. This would mean that all candidates' total capacity, including their motivation, morale, job skills, high-level Japanese (and other language) skills, interpersonal skills, personality, etc., would be assessed and matching of the interests of the organizations and candidates are very important. Some of these companies do not hire foreign professionals. They conduct international dealings instead through overseas companies with which they have concluded joint ventures and licensing agreements, among other kinds of business agreements.

7. The disciplines that the companies most needed personnel to serve their current needs, in order of their priority, according to the results of survey, are:

   a. For *engineering*, (1) electrical engineering, electronics and computers; (2) mechanical engineering; (2) manufacturing engineering; (4) civil engineering; and (5) biotechnology;

   b. For *Science* (which gained about one-third the number of replies of engineering and business), (1) chemistry, (2) Physics, (2) Biology, (4) Medicine, and

   c. For *Business*, (1) accounting, (1) international business, (3) Legal, and (4) others such as industrial art, M & A, loan and securities.

8. Professional qualifications such as certified public accountant (CPA), chartered financial analyst (CFA), certified internal auditor (CIA), and enrolled agent (EA) are treated as neutral or not necessarily preferred by the companies that replied, probably because the work that requires these qualifications are entrusted to specialized companies that permanently employ these professionals. However, registered professional engineers (P. E.s) are preferred by 25% of the companies (mostly, engineering, heavy industry and construction companies) that replied, juris doctors (J. D.s) by 25% of the companies that replied, and Ph. D.s or Sc. D.s in science and engineering by 31.3% of the companies that replied. It must be noted that in the United States the P. E. qualification is required for conducting engineering work and for establishing engineering firms and that engineering documents must be signed by a person with a P. E. qualification to be official. Japanese companies that are subject to legal conflicts with overseas customers and organizations in connection with their products and services replied that they prefer to hire foreign professionals with a J. D. Some of the Japanese companies place a high value on engineers with a Ph. D. or Sc. D. as managers, engineers and

Table 25.2   Degree of Preference for Foreign Professionals Qualifications by Japanese Companies that Replied

Qualification	Degree of Preference
Ph. D. or Sc.D. in Science and Engineering	Preferred by 31.3% of the companies that replied.
Registered Professional Engineer (P. E.)	Preferred by 25% of the companies (in engineering, heavy industry and construction sectors) that replied.
Juris Doctor (J. D.)	Preferred by 25% of the companies that replied.
British quantity surveyor	Preferred by a construction company.
Medical Doctor (M. D.)	Preferred by a pharmaceutical company.
Lawyer (includes J.D.)	Preferred by a computer manufacturing company.
Certified Public Accountant (CPA)	Neutral or Not Preferred
Chartered Financial Analysts (CFA)	Neutral or Not Preferred
Certified Internal Auditor (CIA)	Neutral or Not Preferred
Enrolled Agent (EA)	Neutral or Not Preferred

researchers, since a solid background in these disciplines will highly contribute to the business related to their products and/or services. On the other hand, the larger numbers of Japanese companies still prefer foreign professionals with bachelor's and master's degrees, just as for their Japanese counterparts, as will be stated in 9. Of particular note is that British quantity surveyor is preferred by a construction company, medical doctor by a pharmaceutical company, and a lawyer by a computer manufacturing company, typically reflecting the needs of companies in the respective sectors. Table 25.2 shows the summary of preference on professional qualification by Japanese companies that replied to the questionnaire.

9. Academic qualifications such as bachelor's and master's degrees are preferred by 43.8% of the companies that replied. This typical preference for academic qualifications may be shifted toward a preference for doctoral degrees, in the case of foreign professionals, as expressed by 25% of the companies that replied.

10. Among the companies that replied, 43.8%, 31.3% and 25% of them, respectively, noted that they would prefer to hire foreign professionals via university placement office, advertisements in publications (including Internet), and recommendations from another person. Other replies mentioned hiring via recruiters and head hunting companies.

The trends at national institutes and universities, which tend to offer relatively generous terms and work conditions, may be sensed via other sources. Significant numbers of fellowships are given to foreign professionals, as the current governmental policy emphasizes the internationalization of national institutes and universities. It is also common knowledge that foreign professionals can rarely obtain tenured professorships at national universities in particular, but both national and private universities have begun to offer tenured positions to foreign nationals and this trend is expected to gradually and steadily spread in the future. In cases the author knows of, some European scholars tends to prefer faculty positions in Japan to European and Australian counterparts, especially if they are married to Japanese or Chinese spouses that are based in Japan.

The foreign professional may not be able to behave as freely in Japan as in more open societies, on many occasions given the narrow behavioral boundaries and strict professional practices and ethics typical of this country. Often, very ambitious, conspicuous professionals tend to be kept at respectful distance from a majority of their Japanese professional colleagues. However, once they find the right workplace and are accepted by their Japanese colleagues, the chances are good that they can work in Japan at least for a long time, or, in increasing cases, permanently. Even if these professionals have to leave Japan for some reason, they will likely have opportunities to work at Japanese-affiliated organizations or for organizations that have relationships of any kind with Japanese counterparts, or at any positions that require a broad international horizon.

The Japanese people have learned a great deal from dealing with foreign professionals, whether inside or outside of their organizations, especially for the past decade. A key to the successful employment is matching of the best interests and long-range goals of both foreign professionals and the Japanese organizations they serve, and mutual efforts to jointly work for a stable and satisfactory work environment. Foreign professionals will then know they can enrich their personal lives through **working in Japan.**

Appendix 1
Selected Companies for Engineers and Scientists in Japan

Seclected Companies for Scientists and Engineers in Japan	Address	Phone Number*	Fax Number*
Air Transport			
All Nippon Airways	3-2-5, Kasumigaseki, Chiyoda-ku, Tokyo 100-6027, Japan	(03) 3592-3065	(03) 5756-5679
Japan Airlines	2-4-11, Higashi-Shinagawa, Shinagawa-ku, Tokyo 140-8637, Japan	(03) 5460-3121	(03) 5460-5915
Toa Domestic Airlines/Japan Air System Co.	JAS M1 Bldg. 5-1, Haneda Kuko 3-chome, Ota-ku, Tokyo 144-0041, Japan	(03) 5756-4046	
Automobile			
Honda Motor	2-1-1, Minami-Aoyama, Minato-ku, Tokyo 107-8556, Japan	(03) 3423-1111	(03) 5412-1515
Mazda Motor	3-1, Shinchi, Fuchu-cho, Aki-gun, Hiroshima Pref. 730-8670	(082) 282-1111	(082) 287-5237
Mitsubishi Motors	5-33-8, Shiba, Minato-ku, Tokyo 108-8410 Japan	(03) 3456-1111	(03) 5232-7747
Nissan Motor	6-17-1, Ginza, Chuo-ku, Tokyo 104-8023, Japan	(03) 3543-5523	(03) 3544-0109
Toyota Motor	1, Toyota-cho, Toyota City, Aichi Pref. 471-8571, Japan	(0565) 28-2121	(0565) 23-5708
Yamaha Motor	2500, Shingai, Iwata City, Shizuoka Pref. 438-8501, Japan	(0538) 32-1115	(0538) 37-4252
Chemicals			
Asahi Chemical Industry	1-1-2, Yuraku-cho, Chiyoda-ku, Tokyo 100-8440, Japan	(03) 3507-2730	(03) 3507-2495
Dainippon Ink and Chemicals	3-7-20, Nihonbashi, Chuo-ku, Tokyo 103-8233, Japan	(03) 3272-4511	(03) 3273-7586
Fuji Photo Film	2-26-30, Nishi-Azabu, Minato-ku, Tokyo 106-8620, Japan	(03) 3406-2111	(03) 3406-2193
Hitachi Chemical	2-1-1, Nishi-Shinjuku, Shinjuku-ku, Tokyo 163-0449, Japan	(03) 3346-3111	(03) 5381-3023

*From outside Japan, dial the international access code ("011" in the U.S.) + country code (81) + area code (omit "0"), then rest of number.

Seclected Companies for Scientists and Engineers in Japan	Address	Phone Number*	Fax Number*
Kao	1-14-10, Nihonbashi-Kayabacho, Chuo-ku, Tokyo 103-8210, Japan	(03) 3660-7111	(03) 3660-7044
Konica	1-26-2, Nishi-Shinjuku, Shinjuku-ku, Tokyo 163-0512, Japan	(03) 3349-5251	(03) 3349-5290
Kyowa Hakko Kogyo	1-6-1, Ohtemachi, Chiyoda-ku, Tokyo 100-8185, Japan	(03) 3282-0007	(03) 3284-1968
Mitsubishi Chemical	2-5-2, Marunouchi, Chiyoda-ku, Tokyo 100-0005, Japan	(03) 3283-6254	(03) 3283-6287
Mitsui Chemicals	3-2-5, Kasumigaseki, Chiyoda-ku, Tokyo 100-6070, Japan	(03) 3592-4060	(03) 3592-4213
Sekisui Chemical	2-4-4, Nishi-Tenma, Kita-ku, Osaka 530-8565, Japan	(06) 6365-4122	(06) 6365-4370
Shin-Etsu Chemical	2-6-1, Ohtemachi, Chiyoda-ku, Tokyo 100-0004, Japan	(03) 3246-5011	(03) 3246-5358
Shiseido	7-5-5, Ginza, Chuo-ku, Tokyo 104-8010, Japan	(03) 3572-5111	(03) 3572-6973
Showa Denko	1-13-9, Shiba-Daimon, Minato-ku, Tokyo 105-8518, Japan	(03) 5470-3111	(03) 3431-6442
Sumitomo Chemical	4-5-33, Kitahama, Chuo-ku, Osaka 541-8550, Japan	(06) 6220-3891	(03) 5543-5901
Ube Industries	1-12-32, Nishi-Honmachi, Ube City, Yamaguchi Pref. 755-0052, Japan	(0836) 31-1111	(03) 5460-3390
Communications			
DDI	8, Ichiban-cho, Chiyoda-ku, Tokyo 102-0082, Japan	(03) 3222-0077	(03) 3221-9696
Japan Telecom	4-7-1 Hatcho-bori, Chuo-ku, Tokyo 104-8508, Japan	(03) 5540-8000	(03) 5543-1968
KDD	2-3-2, Nishi-Shinjuku, Shinjuku-ku, Tokyo 163-8003, Japan	(03) 3347-7111	(03) 3347-7000
Nippon Telegraph and Telephone (NTT)	3-19-1, Nishi-Shinjuku, Shinjuku-ku, Tokyo 163-8019, Japan	(03) 5359-5111	(03) 5205-5589
NTT Mobile Communications Network	2-10-1, Toranomon, Minato-ku, Tokyo 105-8436, Japan	(03) 5563-2200	(03) 5572-6646
Construction			
Chiyoda	2-12-1, Tsurumi-Chuo, Tsurumi-ku, Yokohama 230-8601, Japan	(045) 521-1231	(045) 506-9398
JGC	2-2-1, Ohtemachi, Chiyoda-ku, Tokyo 100-0004, Japan	(03) 3279-5441	(045) 682-1112
Obayashi	2-15-2, Konan, Minato-ku, Tokyo 108-8502, Japan	(03) 5769-1017	(03) 5769-1910
Shimizu	1-2-3, Shibaura, Minato-ku, Tokyo 105-8007, Japan	(03) 5441-1111	(03) 5441-0349
Taisei	1-25-1, Nishi-Shinjuku, Shinjuku-ku, Tokyo 163-0606, Japan	(03) 3348-1111	(03) 3345-1386

*From outside Japan, dial the international access code ("011" in the U.S.) + country code (81) + area code (omit "0"), then rest of number.

Seclected Companies for Scientists and Engineers in Japan	Address	Phone Number*	Fax Number*
Toyo Engineering	3-2-5, Kasumigaseki, Chiyoda-ku, Tokyo 100-6005, Japan	(03) 3592-7411	(03) 3593-0749
Electric Power and Gas			
Chubu Electric Power	1, Higashi-Shinmachi, Higashi-ku, Nagoya 461-8680, Japan	(052) 951-8211	(052) 973-3158
Electric Power Development Co.	6-15-1, Ginza, Chuo-ku, Tokyo 104, Japan	(03) 3546-2211	(03) 3546-9531
Kansai Electric Power	3-3-22, Nakanoshima, Kita-ku, Osaka 530-8270, Japan	(06) 6441-8821	(06) 6441-8598
Osaka Gas	4-1-2, Hirano-machi, Chuo-ku, Osaka 541-0046, Japan	(06) 6202-2221	(06) 6226-1681
Tokyo Electric Power	1-1-3, Uchi-Saiwaicho, Chiyoda-ku, Tokyo 100-0011, Japan	(03) 3501-8111	(03) 3592-1795
Tokyo Gas	1-5-20, Kaigan, Minato-ku, Tokyo 105-8527, Japan	(03) 3433-2111	(03) 3437-9190
Electrical Machinery			
Fuji Electric	1-11-2, Ohsaki, Shinagawa-ku, Tokyo 141-0032, Japan	(03) 5435-7111	(03) 5435-7486
Fujitsu	1-6-1, Marunouchi, Chiyoda-ku, Tokyo 100-8211, Japan	(03) 3216-3211	(03) 3216-9365
Hitachi	4-6, Kanda-Surugadai, Chiyoda-ku, Tokyo 101-8010, Japan	(03) 3258-1111	(03) 3258-5480
IBM Japan	2-12, Roppongi 3-chome, Minato-ku, Tokyo 106-8711, Japan	(03) 3586-1111	
Kyocera	6, Takeda-Tobadonocho, Fushimi-ku, Kyoto 612-8501, Japan	(075) 604-3500	(075) 604-3557
Matsushita Electric Industrial	1006, Kadoma, Kadoma City, Osaka Pref. 571-8501, Japan	(06) 6908-1121	(06) 6908-2351
Murata Mfg.	2-26-10, Tenjin, Nagaokakyo City, Kyoto Pref. 617-8555, Japan	(075) 955-6502	(075) 958-2219
NEC	5-7-1, Shiba, Minato-ku, Tokyo 108-8001, Japan	(03) 3454-1111	(03) 3457-7249
Sharp	22-22, Nagaike-cho, Abeno-ku, Osaka 545-8522, Japan	(06) 6621-1221	(06) 6628-1653
Sony	6-7-35, Kita-Shinagawa, Shinagawa-ku, Tokyo 141-0001, Japan	(03) 5448-2111	(03) 5448-2183
Toshiba	1-1-1, Shibaura, Minato-ku, Tokyo 105-8001, Japan	(03) 3457-4511	(03) 5444-9202
Fishery, Agriculture, and Forestry			
Maruha	1-1-2, Ohtemachi, Chiyoda-ku, Tokyo 100-8608, Japan	(03) 3216-0821	(03) 3216-0342
Nippon Suisan	2-6-2, Ohtemachi, Chiyoda-ku, Tokyo 100-8686, Japan	(03) 3244-7000	(03) 3244-7085

*From outside Japan, dial the international access code ("011" in the U.S.) + country code (81) + area code (omit "0"), then rest of number.

Seclected Companies for Scientists and Engineers in Japan	Address	Phone Number*	Fax Number*
Foods			
Ajinomoto	1-15-1, Kyobashi, Chuo-ku, Tokyo 104-8315, Japan	(03) 5254-8111	(03) 5250-8378
Asahi Breweries	1-23-1, Azumabashi, Sumida-ku, Tokyo 130-8602, Japan	(03) 5608-5112	(03) 5608-7121
Ezaki Glico	4-6-5, Utajima, Nishi-Yodogawaku, Osaka 555-8502, Japan	(06) 6477-8351	(06) 6477-5670
Itoham Foods	4-27, Takahata-cho, Nishinomiya City, Hyogo Pref. 663-8586, Japan	(0798) 66-1231	
Japan Tobacco	2-2-1, Toranomon, Minato-ku, Tokyo 105-8422, Japan	(03) 3582-3111	(03) 5572-1441
Kikkoman	339, Noda, Noda City, Chiba Pref. 278-8601, Japan	(0471) 23-5111	(0471) 23-5200
Kirin Breweries	1, Kanda-Izumicho, Chiyoda-ku, Tokyo 101-8645, Japan	(03) 5821-4001	(03) 5821-8455
Meiji Milk Products	2-3-6, Kyobashi, Chuo-ku, Tokyo 104-8381, Japan	(03) 3281-6118	(03) 3281-4717
Meiji Seika	2-4-16, Kyobashi, Chuo-ku, Tokyo 104-8002, Japan	(03) 3272-6511	(03) 3281-7046
Moringa & Co.	5-33-1, Shiba, Minato-ku, Tokyo 108-8403, Japan	(03) 3456-0112	(03) 3769-6129
Moringa Milk Industry	5-33-1, Shiba, Minato-ku, Tokyo 108-8384, Japan	(03) 3798-0111	(03) 3798-0101
Nichirei	6-19-20, Tsukiji, Chuo-ku, Tokyo 104-8402, Japan	(03) 3248-2101	(03) 3248-2119
Nippon Meat Packers	3-6-14, Minami-Honmachi, Chuo-ku, Osaka 541-0054, Japan	(06) 6282-3031	(06) 6282-1056
Nisshin Flour Milling	1-25, Kanda-Nishikicho, Chiyoda-ku, Tokyo 101-8441, Japan	(03) 5282-6666	
Nissin Food Products	4-1-1, Nishi-Nakajima, Yodogawa-ku, Osaka 532-8524, Japan	(06) 6305-7711	(06) 6304-1288
Sapporo Breweries	4-20-1, Ebisu, Shibuya-ku, Tokyo 150-8686, Japan	(03) 5423-2111	(03) 5423-2078
Snow Brand Milk Products	13, Honshio-cho, Shinjuku-ku, Tokyo 160-8575, Japan	(03) 3226-2111	(03) 3226-2150
Takara Shuzo	Karasuma-Higashiiru, Shijodori, Shimogyo-ku, Kyoto 600-8688, Japan	(075) 241-5110	(075) 241-5127
Toyo Suisan	2-13-40, Konan, Minato-ku, Tokyo 108-8501, Japan	(03) 3458-5111	(03) 3450-1381
Yakult Honsha	1-1-19, Higashi-Shinbashi, Minato-ku, Tokyo 105-8660, Japan	(03) 3574-8960	(03) 3575-1636
Yamazaki Baking	3-10-1, Iwamoto-cho, Chiyoda-ku, Tokyo 101-8585, Japan	(03) 3864-3111	(03) 3864-3109

*From outside Japan, dial the international access code ("011" in the U.S.) + country code (81) + area code (omit "0"), then rest of number.

Seclected Companies for Scientists and Engineers in Japan	Address	Phone Number*	Fax Number*
Glass and Ceramics			
Asahi Glass	2-1-2, Marunouchi, Chiyoda-ku, Tokyo 100-8305, Japan	(03) 3218-5555	(03) 3201-5390
Nippon Electric Glass	2-7-1, Seiran, Ohtsu City 520-8639, Japan	(077) 537-1700	(077) 534-4967
Nippon Sheet Glass	2-1-7, Kaigan, Minato-ku, Tokyo 105-8552, Japan	(03) 5443-9522	(03) 5443-9554
Taiheiyo Cement	3-8-1, Nishi-Kanda, Chiyoda-ku, Tokyo 101-8357	(03) 5214-1520	(03) 5214-1707
Land Transport			
Central Japan Railway	2-14-19, Meieki-Minami, Nakamura-ku, Nagoya 450-0003, Japan	(052) 564-2620	(052) 587-1302
East Japan Railway	2-2-2, Yoyogi, Shibuya-ku, Tokyo 151-8578, Japan	(03) 5334-1111	(03) 5334-1106
Hokkaido Railway	1-1, Kita 11-jo, Nishi 15-chome, Chuo-ku, Sapporo 060-8644, Japan	(011) 700-5800	(011) 700-5734
Kinki Nippon Railway	6-1-55, Ue-Honmachi, Tennoji-ku, Osaka 543-8585, Japan	(06) 6775-3444	(06) 6775-3467
Kyushu Railway	1-1, Hakataeki Chuogai, Hakata-ku, Fukuoka 812-8566, Japan	(092) 474-2501	
Nippon Express	3-12-9, Soto-Kanda, Chiyoda-ku, Tokyo 101-8617, Japan	(03) 3253-1111	(03) 5294-5129
Seino Transportation	1, Taguchi-cho, Ohgaki City, Gifu Pref. 503-8501, Japan	(0584) 81-1111	(0584) 82-5045
Shikoku Railway	8-33, Hamano-machi, Takamatsu, Kagawa 760-8580, Japan	(087) 825-1622	
West Japan Railway	2-4-24, Shibata, Kita-ku, Osaka 530-8341, Japan	(06) 6375-8929	(06) 6376-6053
Machinery			
Daikin Industries	2-4-12, Nakazaki-Nishi, Kita-ku, Osaka 530-8323, Japan	(06) 6373-4312	(06) 6373-4330
Ebara	11-1, Haneda-Asahicho, Ohta-ku, Tokyo 144-8510, Japan	(03) 3743-6111	(06) 3745-3010
FANUC	Shibokusa, Oshino-mura, Minami-Tsuru-gun, Yamanashi Pref. 401-0597, Japan	(0555) 84-5555	(0555) 84-5512
Hitachi Construction Machinery	2-6-2, Ohtemachi, Chiyoda-ku, Tokyo 100-0004, Japan	(03) 3245-6305	(03) 3246-2607
Hitachi Zosen	1-7-89, Nankokita, Suminoe-ku, Osaka 559-8559, Japan	(06) 6569-0001	
Komatsu	2-3-6, Akasaka, Minato-ku, Tokyo 107-8414, Japan	(03) 5561-2616	(03) 3505-9662
Koyo Seiko	3-5-8, Minami-Senba, Chuo-ku, Osaka 542-8502, Japan	(06) 6271-8451	(06) 6245-7892

*From outside Japan, dial the international access code ("011" in the U.S.) + country code (81) + area code (omit "0"), then rest of number.

Seclected Companies for Scientists and Engineers in Japan	Address	Phone Number*	Fax Number*
Kubota	1-2-47, Shikitsu-Higashi, Naniwa-ku, Osaka 556-8601, Japan	(06) 6648-2111	(06) 6648-2398
Minebea	1-8-1, Shimo-Meguro, Meguro-ku, Tokyo 153-8662, Japan	(03) 5434-8611	(03) 5434-8603
Mitsubishi Heavy Industries	2-5-1, Marunouchi, Chiyoda-ku, Tokyo 100-8315, Japan	(03) 3212-3111	(03) 3212-9860
NSK	1-6-3, Ohsaki, Shinagawa-ku, Tokyo 141-8560, Japan	(03) 3779-7111	(03) 3779-7445
NTN	1-3-17, Kyomachibori, Nishi-ku, Osaka 550-0003, Japan	(06) 6443-5001	(06) 6443-6966
Sumitomo Heavy Industries	5-9-11, Kita-Shinagawa, Shinagawa-ku, Tokyo 141-8686, Japan	(03) 5488-8000	(03) 5488-8056
Marine Transport			
Kawasaki Kisen	1-2-9, Nishi-Shinbashi, Minato-ku, Tokyo 105-8421, Japan	(03) 3595-5000	(03) 3595-6155
Mistui O.S.K. Lines	2-1-1, Toranomon, Minato-ku, Tokyo 105-8688, Japan	(03) 3587-7111	(03) 3587-7702
Nippon Yusen	2-3-2, Marunouchi, Chiyoda-ku, Tokyo 100-0005, Japan	(03) 3284-5151	(03) 3284-6081
Metal Products			
Tostem	2-1-1, Ohjima, Koto-ku, Tokyo 136-8535, Japan	(03) 3638-8115	(03) 3638-8343
Toyo Seikan	1-3-1, Uchi-Saiwaicho, Chiyoda-ku, Tokyo 100-8522, Japan	(03) 3508-2113	
Mining			
Mitsui Mining	2-1-1, Nihonbashi-Muromachi, Chuo-ku Tokyo 103-0022, Japan	(03) 3241-1334	(03) 3241-8684
Sumitomo Coal Mining	3-20-4, Nishi-Shinbashi, Minato-ku, Tokyo 105-8425	(03) 5404-0401	(03) 5404-0445
Nonferrous Metals			
Fujikura	1-5-1, Kiba, Koto-ku, Tokyo 135-8512, Japan	(03) 5606-1030	(03) 5606-1502
Furukawa Electric	2-6-1, Marunouchi, Chiyoda-ku, Tokyo 100-8322, Japan	(03) 3286-3001	(03) 3286-3694
Mitsubishi Materials	1-5-1, Ohtemachi, Chiyoda-ku, Tokyo 100-8117, Japan	(03) 5252-5201	(03) 5252-5281
Mitsui Mining & Smelting	1-11-1, Ohsaki, Shinagawa-ku, Tokyo 141-8584, Japan	(03) 5437-8000	(03) 5437-8033
Nippon Light Metal	2-2-20, Higashi-Shinagawa, Shinagawa-ku, Tokyo 140-8628, Japan	(03) 5461-9211	(03) 5461-9344

*From outside Japan, dial the international access code ("011" in the U.S.) + country code (81) + area code (omit "0"), then rest of number.

Seclected Companies for Scientists and Engineers in Japan	Address	Phone Number*	Fax Number*
Showa Aluminum	3-6-5, Iidabashi, Chiyoda-ku, Tokyo 102-8111, Japan	(03) 3239-5311	(03) 3239-5306
Sumitomo Electric Industries	4-5-33, Kitahama, Chuo-ku, Osaka 541-0041, Japan	(06) 6220-4141	(06) 6222-6478
Sumitomo Metal Mining	5-11-3, Shinbashi, Minato-ku, Tokyo 105-8716, Japan	(03) 3436-7704	(03) 3436-7735
Oil and Coal Products			
Cosmo Oil	1-1-1, Shibaura, Minato-ku, Tokyo 105-8528, Japan	(03) 3798-3211	(03) 3798-3411
Idemitsu Kosan Co., Ltd.	1-1, Marunouchi 3-chome, Chiyoda-ku, Tokyo, Japan	(03) 3213-9330	(03) 3213-8087
Japan Energy	2-10-1, Toranomon, Minato-ku, Tokyo 105-8407, Japan	(03) 5573-6000	(03) 5573-6773
Nippon Mitsubishi Oil	1-3-12, Nishi-Shinbashi, Minato-ku, Tokyo 105-8412, Japan	(03) 3502-1135	(03) 3502-9352
Showa Shell Sekiyu	2-3-2, Daiba, Minato-ku, Tokyo 135-8074, Japan	(03) 5531-5601	(03) 5531-5609
Pharmaceuticals			
Daiichi Pharmaceutical	3-14-10, Nihonbashi, Chuo-ku, Tokyo 103-8234, Japan	(03) 3272-0611	(03) 3281-8427
Eisai	4-6-10, Koishikawa, Bunkyo-ku, Tokyo 112-8088, Japan	(03) 3817-3700	(03) 3811-3077
Takeda Chemical Industries	4-1-1, Dosho-machi, Chuo-ku, Osaka 540-8645, Japan	(06) 6204-2111	(06) 6204-2035
Yamanouchi Pharmaceutical	2-3-11, Nihonbashi, Chuo-ku, Tokyo 103-8411, Japan	(03) 3244-3000	(03) 5255-7662
Precision Instruments			
Canon	3-30-2, Shimomaruko, Ohta-ku, Tokyo 146-8501, Japan	(03) 3758-2111	(03) 5482-5135
Citizen Watch	2-1-1, Nishi-Shinjuku, Shinjuku-ku, Tokyo 163-0428, Japan	(03) 3342-1231	(03) 3342-1280
Dainippon Screen Mfg.	4-1-1, Teranouchi-Agaru, Horikawadori, Kamigyo-ku, Kyoto 602-8585, Japan	(075) 414-7111	(075) 414-7100
Fuji Zerox	2-17-22 Akasaka, Minato-ku, Tokyo 107-0052, Japan www.fujixerox.co.jp/	(03) 3585-3211	
Minolta	2-3-13, Azuchi-machi, Chuo-ku, Osaka 541-8556, Japan	(06) 6271-2251	(06) 6271-8320
Nikon	3-2-3, Marunouchi, Chiyoda-ku, Tokyo 100-8331, Japan	(03) 3214-5311	(03) 3216-1454
Olympus Optical	2-3-1, Nishi-Shinjuku, Shinjuku-ku, Tokyo 163-0914	(03) 4430-2111	(03) 3340-2098
Ricoh	1-15-5, Minami-Aoyama, Minato-ku, Tokyo 107-8544, Japan	(03) 3479-3111	(03) 3403-1578

*From outside Japan, dial the international access code ("011" in the U.S.) + country code (81) + area code (omit "0"), then rest of number.

Seclected Companies for Scientists and Engineers in Japan	Address	Phone Number*	Fax Number*
Seiko	2-6-21, Kyobashi, Chuo-ku, Tokyo 104-8331, Japan	(03) 3563-2111	(03) 3563-8496
Shimadzu	1, Nishinokyo-Kuwabaracho, Nakagyo-ku, Kyoto 604-8511, Japan	(075) 823-1111	(075) 822-0709
Pulp and Paper			
Nippon Paper Industries	1-12-1, Yuraku-cho, Chiyoda-ku, Tokyo 100-0006, Japan	(03) 3218-8000	(03) 3214-5226
Oji Paper	4-7-5, Ginza, Chuo-ku, Tokyo 104-0061, Japan	(03) 3563-1111	(03) 3563-1132
Rengo	2-5-25, Umeda, Kita-ku, Osaka 530-0001, Japan	(06) 6345-2371	(06) 6342-0376
Rubber Products			
Bridgestone	1-10-1, Kyobashi, Chuo-ku, Tokyo 104-8340, Japan	(03) 3567-0111	(03) 3567-4615
Sumitomo Rubber Industries	3-6-9, Wakihama-cho, Chuo-ku, Kobe 651-0072, Japan	(078) 265-3000	(078) 265-3113
Yokohama Rubber	5-36-11, Shinbashi, Minato-ku, Tokyo 105-8685, Japan	(03) 3432-7111	(03) 3432-5616
Services			
Benesse	3-7-17, Minamigata, Okayama City 700-8686, Japan	(086) 225-1100	(042) 356-7302
CSK	2-6-1, Nishi-Shinjuku, Shinjuku-ku, Tokyo 163-0227, Japan	(03) 3344-1811	(03) 5321-3939
Hitachi Information Systems	1-16-5, Dogenzaka, Shibuya-ku, Tokyo 150-8540, Japan	(03) 3464-5110	(03) 3496-5684
INTEC	5-5, Ushijima-Shinmachi, Toyama City 930-8577, Japan	(0764) 44-1111	(0764) 44-8015
KONAMI	4-3-1, Toranomon, Minato-ku, Tokyo 105-6021, Japan	(03) 3432-5678	(03) 3432-5679
NTT Data	3-3-3, Toyosu, Koto-ku, Tokyo 135-6033, Japan	(03) 5546-8202	(03) 5546-8145
SECOM	1-26-2, Nishi-Shinjuku, Shinjuku-ku, Tokyo 160-0555, Japan	(03) 3348-7511	(03) 3345-0219
Softbank	24-1, Nihonbashi-Hakozakicho, Chuo-ku, Tokyo 103-8501, Japan	(03) 5642-8000	(03) 5641-3401
Steel Products			
Kawasaki Steel	2-2-3, Uchi-Saiwaicho, Chiyoda-ku, Tokyo 100-0011, Japan	(03) 3597-3111	(03) 3597-4911
Kobe Steel	2-10-26, Wakinohama-cho, Chuo-ku, Kobe 651-8585, Japan	(078) 261-5111	(03) 5252-7961
Nippon Steel	2-6-3, Ohtemachi, Chiyoda-ku, Tokyo 100-8071, Japan	(03) 3242-4111	(03) 3275-5611
NKK	1-1-2, Marunouchi, Chiyoda-ku, Tokyo 100-8202, Japan	(03) 3212-7111	(03) 3214-8436

*From outside Japan, dial the international access code ("011" in the U.S.) + country code (81) + area code (omit "0"), then rest of number.

Seclected Companies for Scientists and Engineers in Japan	Address	Phone Number*	Fax Number*
Sumitomo Metals	4-5-33, Kitahama, Chuo-ku, Osaka 541-0041, Japan	(06) 6220-5111	(06) 6223-0563
Textiles and Apparel			
Kanebo	3-20-20, Kaigan, Minato-ku, Tokyo 108-8080, Japan	(03) 5446-3002	(03) 5446-3027
Kuraray	1-12-39, Umeda, Kita-ku, Osaka 530-8611, Japan	(06) 6348-2111	(06) 6348-2165
Teijin	1-6-7, Minami-Honmachi, Chuo-ku, Osaka 541-8587, Japan	(06) 6268-2132	(06) 6268-3205
Toray Industries	2-2-1, Nihonbashi-Muromachi, Chuo-ku, Tokyo 103-8666, Japan	(03) 3245-5111	(03) 3245-5459
Toyobo	2-2-8, Dojimahama, Kita-ku, Osaka 530-8230, Japan	(06) 6348-3111	(06) 6348-3192
Transport Equipment			
Ishikawajima-Harima Heavy Industries (IHI)	2-2-1, Ohtemachi, Chiyoda-ku, Tokyo 100-8182, Japan	(03) 3244-5111	(03) 3244-5139
Kawasaki Heavy Industries	1-1-3, Higashi-Kawasakicho, Chuo-ku, Kobe 650-8680, Japan	(078) 371-9530	(03) 3432-4759
Mitsui Engineering & Shipbuilding	5-6-4, Tsukiji, Chuo-ku, Tokyo 104-8439, Japan	(03) 3544-3147	(03) 3544-3050
Other Products			
Dai Nippon Printing	1-1-1, Ichigaya-Kagacho, Shinjuku-ku, Tokyo 162-8001, Japan	(03) 3266-2111	(03) 3266-2129
Nintendo	60, Kami-Takamatsucho, Fukuine, Higashiyama-ku, Kyoto 605-8660	(075) 541-6111	
Sega Enterprises	1-2-12, Haneda, Ohta-ku, Tokyo 144-8531, Japan	(03) 5736-7111	(03) 5736-7058
Shin-Etsu Polymer	4-3-5, Nihonbashi-Honcho, Chuo-ku, Tokyo 103-0023, Japan	(03) 3279-1712	(03) 3246-2529
Toppan Printing	1, Kanda-Izumicho, Chiyoda-ku, Tokyo 101-0024, Japan	(03) 3835-5111	(03) 3837-7675
Yamaha	10-1, Nakazawa-machi, Hamamatsu City, Shizuoka Pref. 430-8560, Japan	(053) 460-2071	(053) 464-8554

*From outside Japan, dial the international access code ("011" in the U.S.) + country code (81) + area code (omit "0"), then rest of number.

Appendix 2

National Laboratories and Research Organizations in Japan

National Laboratories and Research Organizations in Japan	Address	Phone Number*	Fax Number*	Typical Research Activities and Web Address (If Available)
Agricultural Policy Research Committee, Inc. (APRC)	26-3, Nishigahara 1-chome, Kita-ku, Tokyo 114, Japan	(03) 3910-7223	(03) 3910-7267	Vital agriculture and food problems, and agricultural policies
Akajima Marine Science Laboratory (AMSL)–Establishment of Tropical Marine Ecological Research (ETMER)	Tokyo Office: 26-2, Nishi-gotanda 1-chome, Shinagawa-ku, Tokyo 141, Japan AMSL: 179, Aka, Zamami-mura, Shimajiri-gun, Okinawa Pref. 901-33, Japan	(03) 3490-7266 (098) 987-2304	(03) 3490-8278 (098) 987-2875	Sexual reproduction and larval ecology of scleractinian corals, F/S for the development of aquaculture and sea farming of tropical and subtropical invertebrates
Aomori Advanced Industrial Technology Center	202-4, Ashiya, Yatsuyaku, Aomori City, Aomori Pref., 030-01, Japan	(0177) 39-9676	(0177) 39-9613	Web:www.aomori-tech.go.jp Biotechnology mechanics and electronics
Building Research Institute (BRI)	1, Tachihara Tsukuba City, Ibaraki Pref, 305, Japan	(0298) 64-2151	(0298) 64-2989	Web: www.kenken.go.jp Construction methods
Central Customs Laboratory (CCL)	531, Iwase, Matsudo City, Chiba Pref. 271, Japan	(047) 363-4211	(047) 361-0531	Detection instruments for controlled drugs
Central Research Institute of the Electric Power Industry (CRIEPI)	1-6-1 Otemachi, Chiyoda-ku, Tokyo 100, Japan	(03) 3201-6601	(03) 3287-2841	Web:www.ic-ml@criepi.Denken.or.jp Nuclear and fossil fuel power generation, the environment
Chemicals Inspection and Testing Institute, Japan	5-6-21, Kameido, Koto-ward, Tokyo 136, Japan	(03) 3638-8847	(03) 3638-8840	High-sensitivity methods for trace elements, biodegradability

Organization	Address			Research focus
Chugoku National Agricultural Experiment Station	6-12-1, Nishifukatsu, Fukuyama City, Hiroshima Pref. 721, Japan	(0849) 23-4100	(0849) 24-7893	Farming systems, high-value production of main crops
Chugoku National Industrial Research Institute	2-2-2, Hiro-Suehiro, Kure City, Hiroshima Pref. 737-01, Japan	(0823) 72-1111	(0823) 73-3284	Materials science, computer science, biotechnology
Civil Engineering Research Institute, Hokkaido Development Bureau	Hiragishi 1-jo 3-chome, Toyohiraku, Sapporo City, Hokkaido 062, Japan	(011) 841-1111	(011) 824-1226	River and port study, structures, road, agricultural engineering
Communications Research Laboratory (CRL)	4-2-1 Nukui-Kitamachi, Koganei, Tokyo 184, Japan	(0423) 21-1211	(0423) 27-7458	Highly intelligent communications, human informatics
Electronic Navigation Research Institute (ENRI)	6-38-1, Shinkawa, Mitaka-City, Tokyo 181, Japan	(0422) 41-3162	(0422) 41-3169	Safety of traffic systems, prevention of pollution
Electrotechnical Laboratory (ETL)	1-1-4 Umezono, Tsukubashi, Ibaraki 305, Japan	(0298) 54-5006 (Research Planning Office); (0298) 54-5013 (International Research Cooperation Office)	(0298) 54-5087	Basic electronics, standards and measurements, energy-related technology, information technology, X-ray laser, synchrotron radiation applications
Forestry and Forest Products Research Institute (FFPRI)	PO Box 16, Tsukuba Norin Kenkyu Danchi-nai, Ibaraki Pref. 305, Japan	(0298) 73-3211	(0298) 74-3720 or 74-8507	Forest-ecosystem properties, forest resources
Fukuoka Industrial Technology Center (FITC)	332-1 Kamikoga, Chikushino, Fukuoka Pref. 818, Japan	(092) 925-7721	(092) 925-7724	Surveillance and control systems, recycling technology
Geographical Survey Institute (GSI)	Kitasato-1, Tsukuba City, Ibaraki Pref. 305, Japan	(0298) 64-1111	(0298) 64-8087	Establishment of geodetic control points
Geographical Survey of Japan (GSI)	1-1-3 Higashi, Tsukuba City, Ibaraki Pref. 305, Japan	(0298) 54-3576	(0298) 54-3533 or 56-4989	Geological mapping and compilation of the Japanese islands
Hokkaido National Agricultural Experiment Station	Hitsujigaoka, Toyohira-ku, Sapporo City 062, Japan	(011) 851-9141	(011) 859-2178	Rice culture, upland farming, horticulture, grassland agriculture

* From outside Japan, dial the international access code ("011" in the U.S.) + country code (81) + area code (omit "0"), then rest of number.

Source: Constructed from the information contained in National Laboratories and Public Research Organizations in Japan, published by JST.

National Laboratories and Research Organizations in Japan	Address	Phone Number*	Fax Number*	Typical Research Activities and Web Address (If Available)
Hokkaido National Fisheries Research Institute	116 Katsurakoi, Kushiro, Hokkaido 085, Japan	(0154) 91-9136	(0154) 91-9355	Resource management and enhancement, fisheries oceanography
Hokkaido National Industrial Research Institute (HNRI)	2-17-2-1 Tsukisamu-Higashi, Toyohira-ku, Sapporo 062, Japan	(011) 857-8400	(011) 857-8900	Microgravity research, low-temperature engineering
Hokkaido River Disaster Prevention Research Center	2nd Yuraku Building, West 1, South 1, Chuo-ku, Sapporo City, Hokkaido 060, Japan	(011) 222-8141	(011) 231-3380	Prevention techniques, public hearings, river information, flood control facilities
Hokuriku National Agricultural Experiment Station	Inada, Joetsu 943-01, Niigata, Japan	(0255) 23-4131	(0255) 24-8578	Rice agricultural development, genetics, biotechnology
Hydrographic Department, Maritime Safety Agency	3-1 Tsukiji 5-chome, Chuo-ku, Tokyo 104, Japan	(03) 3541-3813	(03) 3545-2885	Web: www.jhd.go.jp Diffusion in the deep ocean, sea bottom movement study
Industrial Research Center of Ehime Prefecture	487-2 Kumekubota-machi, Matsuyama City, Ehime Pref. 791-11, Japan	(089) 976-7612	(089) 976-7313	Web: www.iri.pref.ehime.jp New metallic materials, industrial robots
Industrial Technology Center of Okayama Prefecture	5301, Haga, Okayama City, Okayama Pref. 701-12, Japan	(086) 286-9600	(086) 286-9630	Materials and systems engineering, development of textiles
Institute for Environmental Sciences	1-7 Ienomae, Obuchi, Rokkashi-mura, Kamikita-gun, Aomori Pref. 039-32, Japan	(0175) 71-1200	(0175) 72-3690	Surveys on characteris-tics of local environment, radiation, radioactivity
Institute for Sea Training	5-57 Kitanaka Dori, Naka-ku, Yokohama 231, Japan	(045) 211-7313	(045) 211-7317	Ship-related studies, marine pollution
Institute of Agricultural Machinery (IAM)–Bio-oriented Technology Research Advancement Institution (BRAIN)	1-40-2 Nisshin Omiya, Saitama Pref. 331, Japan	(048) 663-3901-4	(048) 651-9655	Agricultural automation, biotechnology engineering, ergonomics, machine dynamics and reliability, crop production machinery and systems
Institute of Research and Innovation (IRI)	1-6-8 Yushima, Bunkyo-ku, Tokyo 113, Japan	(03) 5689-6351	(03) 5689-6350	Energy- and environment-related researches

Organization	Address	Phone	Fax	Research
International Association of Traffic and Safety Sciences (IATSS)	6-20, 2-chome, Yaesu, Chuo-ku, Tokyo 104, Japan	(03) 3273-7884	(03) 3272-7054	Research Related to the Aged Traffic Safety Systems, Transportation Mode Choices, Communication Systems
International Development Center of Japan (IDCJ)	Kyofuku Building, 9-11, Tomioka 2-chome, Koto-ku, Tokyo 135, Japan	(03) 3630-6911	(03) 3630-8120	Country and regional studies, sector studies, aid management
Iwate Biotechnology Research Center (IBRC)	Narita 22-174-4, Kitakami City, Iwate 024, Japan	(0197) 68-2911	(0197) 68-3881	New cultivar mother plants, disease resis-tant parental plants
Iwate Industrial Research Institute (IIRI)	3-35-2, Iiokahinden, Morioka City, Iwate Pref. 020, Japan	(019) 635-1115	(019) 635-0311	Web: www.kiri.pref.iwate.jp/kiri 3-D computer tomography applications
Japan Atomic Energy Research Institute (JAERI)	Fukoku Seimei Bldg., 2-2-2 Uchisaiwai-cho, Chiyoda-ku, Tokyo 100, Japan	(03) 3592-2111	(03) 3580-6107	Web: www.jaeri.go.jp Comprehensive R&D on nuclear energy
Japan Automobile Research Institute (JARI)	Karima 2530, Tsukuba City, Ibaraki Pref. 305, Japan	(0298) 56-1111	(0298) 56-1122	Safety, environment, fuels/lubricants, new technologies
Japan Chemical Analysis Center (JCAC)	295-3, Sanno-cho, Inage-ku, Chiba City, Chiba Pref. 263, Japan	(043) 423-5325	(043) 423-5326	Environmental radioactivity analysis, instrumental analysis
Japan Environmental Sanitation Center (JESC)	10-6, Yotsuyakamicho, Kawasaki-ku, Kawasaki City, Kanagawa Pref. 210, Japan	(044) 288-4896	(044) 299-2294	Public pollution control and pest control, just plan of waste disposal
Japan Fine Ceramics Center (JFCC)	Administrative Department, 4-1 Mutsuno 2-chome, Atsuta -ku, Nagoya 456, Japan	(052) 871-3500	(052) 871-3599	R&D on test, evaluation, manufacturing, application, etc., for fine ceramics
Japan International Research Center for Agricultural Sciences (JIRCAS)	Ohwashi 1-2, Tsukuba City, Ibaraki Pref. 305, Japan	(0298) 38-6313	(0298) 38-6316	Surveys on Agriculture, Forestry, Fisheries, Crop Production, etc., in the Developing Regions.
Japan Marine Science & Technology Center (JAMSTEC)	2-5 Natsushima-cho, Yokosuka City, Kanagawa Pref. 237, Japan	(0468) 66-3811	(0468) 66-2119	Web: www.jamstec.go.jp Comprehensive research on marine science and technology

* From outside Japan, dial the international access code ("011" in the U.S.) + country code (81) + area code (omit "0"), then rest of number.

Source: Constructed from the information contained in National Laboratories and Public Research Organizations in Japan, published by JST.

National Laboratories and Research Organizations in Japan	Address	Phone Number*	Fax Number*	Typical Research Activities and Web Address (If Available)
Japan Science and Technology Corporation (JST)	Kawaguchi Center Building, 1-8 Honcho 4-chome, Kawaguchi City, Saitama Pref. 332, Japan	(048) 226-5630	(048) 226-5652	Web: www.jst.go.jp Administration of basic research, cooperation, technology transfer
Japan Sea National Fisheries Research Institute	Suido-cho, Niigata City, Niigata Pref. 951, Japan	(025) 228-0451	(025) 224-0950	Ecological research on pelagic and ground fish, aquatic organisms
Japan Sewage Works Agency (JS)	Shimosasame 5141, Toda City, Saitama Pref. 335, Japan	(048) 421-2693	(048) 421-7542	Waste water treatment, sewage sludge handling and disposal
Japan Synchrotron Radiation Research Institute	1503-1, Kanaji, Kamigori-cho, Ako-gun, Hyogo Pref. 678-12, Japan	(07915) 8-0960	(07915) 8-0965	R&D on storage ring, basic beam line technologies and utilization
Japan Weather Association (JWA)	Sunshine 60, 3-1-1 Higashi-Ikebukuro, Toshima-ku, Tokyo 170, Japan	(03) 5958-8161	(03) 5958-8162	Meteorological observations and investigations, forecasting services, remote sensing
Japan Wildlife Research Center (JWRC)	2-29-3 Yushima, Bunkyo-ku, Tokyo 113, Japan	(03) 3813-8806	(03) 3813-8958	Researches on wildlife species, management of wildlife and habitat
Japanese Foundation for Cancer Research	Kami-Ikebukuro 1-37-1, Toshima-ku, Tokyo 170, Japan	(03) 3918-0111	(03) 3918-0167	Regulatory mechanisms of cytokines, human T cell leukemia virus
Kanagawa Academy of Science and Technology (KAST)	Kanagawa Science Park, 3-2-1 Sakado, Takatsu-ku, Kawasaki-shi, Kanagawa Prefecture 213, Japan	(044) 819-2030	(044) 819-2026	Ligand/receptor interaction, photon control, photochemical conversion materials
Kanagawa Cancer Center Research Institute	1-2 Nakao 1-chome, Asahi-ku, Yokohama city, Pref. 241, Japan	(045) 391-5761	(045) 366-3157	Pathology, cytogenitics, biochemistry, molecular cell biology
Kanagawa Industrial Technology Research Institute (KITRI)	705-1 Shimoimaizumi, Ebina City, Kanagawa Pref. 243-04, Japan	(0462) 36-1500	(0462) 36-1526	Web: www.kanagawa-iri.go.jp Ultrafine particles, michromachining, safety assessment
Kyushu National Agricultural Experiment Station	2421 Nishigoshi, Kikuchi, Kumamoto Pref. 861-11, Japan	(096) 242-1150	(096) 249-1002	Integration of Agricultural Technologies, Crop Culti-vation Technology

Institution	Address	Phone	Fax	Research
Kyushu National Industrial Research Institute	Shuku-machi, Tosu City, Saga Pref. 841, Japan	(0942) 82-5161	(0942) 83-0850	New carbon, metal, ceramics, polymer, coal liquefaction processes
Lake Biwa Research Institute	1-10 Uchidehama, Otsu, Shiga 520, Japan	(0775) 26-4800	(0775) 26-4803	Natural, social, and cultural environment of lake and catchment area
Marine Technical College (MTC)	12-24 Nishikura-cho, Ashiya, Hyogo Pref. 659, Japan	(0797) 22-9341	(0797) 32-7904	Radio navigation, bridge simulator, radar simulator
Mechanical Engineering Laboratory (MEL)	Namiki 1-2, Tsukuba City, Ibaraki Pref. 305, Japan	(0298) 58-7016	(0298) 58-7033	Web: www.aist.gp.jp/MEL/ Information and systems science, bioengineering, mechanics, design
Meteorological Research Institute (MRI)	1-1, Nagamine, Tsukuba City, Ibaraki Pref. 305, Japan	(0298) 53-8535	(0298) 53-8545	Forest, climate, typhoon, satellite and observation, physical meteorology, seismology
Nansei National Fisheries Research Institute	17-5, Maruishi 2-chome, Ohno-cho, Saeki-gun, Hiroshima Pref. 739-04, Japan	(0829) 55-0666	(0829) 54-1216	Forecast and management of environmental change and marine resources and cultivation
National Aerospace Laboratory (NAL)	7-44-1 Jindaiji-higashimachi, Chofu City, Tokyo 182, Japan	(0422) 47-5911	(0422) 48-5888	Supersonic transport, air safety and environment, H-II orbiting plane, Spaceplane, space propulsion systems
National Agriculture Research Center (NARC)	3-1-1 Kan-nondai, Tsukuba City, Ibaraki Pref. 305, Japan	(0298) 38-8481	(0298) 38-8484	Agricultural production systems, farmland utilization, diseases
National Cancer Center Research Institute	1-1 Tsukiji 5-chome, Chuo-ku, Tokyo 104, Japan	(03) 3542-2511	(03) 3545-3567	Primary prevention, environmental mutagens-carcinogens
National Cardiovascular Center Research Institute	7-1 Fujishiro-dai, 5-chome, Suita City, Osaka 565, Japan	(06) 833-5012	(06) 872-7485	Cellular and molecular aspects of mechanisms, structure, functions
National Children's Medical Research Center	3-35-31 Taishido, Setagaya-ku, Tokyo 154, Japan	(03) 3414-8121	(03) 3414-3208	Cell differentiation and developmental processes, immunology

* From outside Japan, dial the international access code ("011" in the U.S.) + country code (81) + area code (omit "0"), then rest of number.

Source: Constructed from the information contained in National Laboratories and Public Research Organizations in Japan, published by JST.

National Laboratories and Research Organizations in Japan	Address	Phone Number*	Fax Number*	Typical Research Activities and Web Address (If Available)
National Food Research Institute (NFRI)	2-1-2 Kannondai, Tsukuba City, Ibaraki Pref. 305, Japan	(0298) 38-7971	(0298) 38-7996	Web: ss.nfri.affrc.go.jp Food science, analysis, assessment and function, microbiology
National Grassland Research Institute (NGRI)	Nishinasuno, Nasu-gun, Tochigi Pref. 329-27, Japan	(0287) 36-0111	(0287) 36-6629	Web: ss.ngri.affrc.go.jp Production of forage plants, ecosystem
National Industrial Research Institute of Nagoya (NIRIN)	Hirate-cho 1-chome, Kita-ku, Nagoya City, Aichi Pref. 462, Japan	(052) 911-2111	(052) 916-2802	Synergy ceramics, superconducting materials, solar and bioenergy
National Institute for Advanced Interdisciplinary Research (NAIR)	1-1-4 Higashi, Tsukuba City, Ibaraki Pref. 305, Japan	(0298) 54-2500	(0298) 54-2524	Web: www.aist.go.jp/NAIR Atom-molecule manipulation, cluster science, bionic design
National Institute for Environmental Studies (NIES)	16-2 Onogawa, Tsukuba City, Ibaraki Pref. 305, Japan	(0298) 50-2308	(0298) 51-4732	Natural, social, and health sciences, global environmental research
National Institute for Minamata Disease	3-7-1 Kagamiyama, Higasihiroshima-shi, Hiroshima 739, Japan	(0824) 20-0800	(0824) 20-0802	Studies on mercury poisoning incidences, diagnosis, clinical course, etc., of disease
National Institute for Research in Inorganic Materials (NIRIM)	1-1, Namiki, Tsukuba City, Ibaraki Pref. 305, Japan	(0298) 51-3351	(0298) 52-7449	Characterization, synthesis, and property analysis of refractory, electronic materials
National Institute for Resources and Environment (NIRE)	16-3 Onogawa, Tsukuba, Ibaraki 305, Japan	(0298) 58-8100, 8111	(0298) 58-8118	Minimum environmental impact, maximum energy and resource utilization
National Institute of Agro-Environmental Sciences (NIAES)	PO Box 2, Norin Danchi, Tsukuba City, Ibaraki Pref. 305, Japan	(0298) 38-8145, 8161	(0298) 38-8199	Conservation of agro-environmental resources, agro-ecosystems
National Institute of Agrobiological Resources (NIAR)	2-1-2, Kannondai, Tsukuba City, Ibaraki Pref. 305, Japan	(0298) 38-7406	(0298) 38-7408	Research on genetic resources such as seeds and plant tissue, molecular genetics, biotechnology, plant physiology, radiation breeding
Institute of Radiation Breeding:	Ohmiya-machi, Naka-gun, Ibaraki Pref. 319-22, Japan	(0298) 2-1138	(0298) 3-1075	

National Institute of Animal Health (NIAH)	Kannondai 3-1-1, Tsukuba City, Ibaraki Pref. 305, Japan	(0298) 38-7707	(0298) 38-7880, 7907	Pathogenic agents, infectious diseases, immunology
National Institute of Animal Industry	2, Ikenodai, Kukisaki-machi, Inashiki-gun, Ibaraki Pref. 305, Japan	(0298) 38-8600	(0298) 38-8606	Genetic mechanisms and breeding, physiological functions
National Institute of Bioscience and Human Technology (NIBH)	1-1 Higashi, Tsukuba City, Ibaraki Pref. 305, Japan	(0298) 54-6023	(0298) 54-6005	Protein molecular assemblies, complex carbohydrates, medical and health apparatus
National Institute of Fruit Tree Science	2-1 Fujimoto, Tsukuba City, Ibaraki Pref. 305, Japan	(0298) 38-6416	(0298) 38-6437	Breeding, pomology, plant protection: apples, citriculture, grapes
National Institute of Health Sciences (NIHS)	1-18-1 Kamiyoga, Setagaya-ku, Tokyo 158, Japan	(03) 3700-1141	(03) 3707-6950	Physiochemical and biological testing and research on drugs
National Institute of Health Services Management (NIHSM)	1-23-1 Toyama, Shinjuku-ku, Tokyo 162, Japan	(03) 3203-5327	(03) 3202-6853	Health care policy, health care economics, physical design and planning, education
National Institute of Industrial Health	21-1 Nagao 6-chome, Tama-ku, Kawasaki City, Kanagawa Pref. 214, Japan	(044) 865-6111	(044) 865-6116	Work stress, workers' health, occupational diseases
National Institute of Industrial Safety (NIIS)	Umezono 1-4-6, Kiyose, Tokyo 204, Japan	(0424) 91-4512	(0424) 91-7846	Web: www.anken.go.jp Equipment safety, component failure
National Institute of Infectious Diseases (NIID)	23-1 Toyama 1-chome, Shinjuku-ku, Tokyo 162, Japan	(03) 5285-1111	(03) 5285-1150	Causative agents and pathogenesis, diagnosis, potency assay tests
National Institute of Materials and Chemical Research (NIMC)	1-1 Higashi, Tsukuba, Ibaraki Pref. 305, Japan	(0298) 54-6227	(0298) 54-6233	Organic, inorganic, polymeric, and composite materials, energy conservation

* From outside Japan, dial the international access code ("011" in the U.S.) + country code (81) + area code (omit "0"), then rest of number.

Source: Constructed from the information contained in National Laboratories and Public Research Organizations in Japan, published by JST.

National Laboratories and Research Organizations in Japan	Address	Phone Number*	Fax Number*	Typical Research Activities and Web Address (If Available)
National Institute of Mental Health, National Center of Neurology and Psychiatry (NCNP)	1-7-3, Kohnodai, Ichikawa City, Chiba Pref. 272, Japan	(047) 372-0141	(047) 371-2900	Community mental health and care, drug dependence, mind-body correlation and psycho-somatic disorders
National Institute of Neuroscience, NIN	4-1-1 Ogawahigashi, Kodaira City, Tokyo 187, Japan	(0423) 41-2711	(0423) 44-6745	Research on nervous, mental, muscular, and developmental disorders
National Institute of Population and Social Security Research	2-3 Kasumigaseki 1-chome, Chiyoda-ku, Tokyo 100, Japan	(03) 3595-2984		Theoretical and empirical social security, population dynamics
National Institute of Public Health (NIPH)	4-6-1 Shirokanedai, Minato-ku, Tokyo 108, Japan	(03) 3441-7111	(03) 3446-4314	Health policy, promotion and management, environmental health
National Institute of Radiological Sciences (NIRS)	9-1 Anagawa 4-chome, Inage-ku, Chiba City, Chiba Pref. 263, Japan	(043) 251-2111	(043) 256-9616	Physics and chemistry, biomedicine, clinical medicine, environment
National Institute of Science and Technology Policy (NISTEP)	1-11-39 Nagata-cho, Chiyoda-ku, Tokyo 100, Japan	(03) 3581-2391	(03) 3503-3996	Web: www.nistep.go.jp Dynamic analysis, macroscopic structure, etc., of S&T activities
National Institute of Sericultural and Entomological Science (NISES)	Ohwashi 1-2, Tsukuba City, Ibaraki Pref. 305, Japan	(0298) 38-6026	(0298) 38-6028	Gene expression and biological mechanisms of insects, chemical and physical analysis
National Research Institute for Earth Science and Disaster Prevention (NIED)	3-1 Tennodai, Tsukuba City, Ibaraki Pref. 305, Japan	(0298) 51-1611	(0298) 51-1622	Atmospheric, hydrospheric, solid earth and seismic sciences, disaster prevention research
National Research Institute for Metals (NRIM)	1-2-1 Sengen, Tsukuba City, Ibaraki Pref. 305, Japan	(0298) 59-2000	(0298) 59-2049	Superconductors, heat-resisting alloys, intermetallic compounds
National Research Institute of Agricultural Economics	2-1 Nishigahara 2-chome, Kita-ku, Tokyo 114, Japan	(03) 3910-3946	(03) 3940-0232	Economic analysis, structural analysis for high productivity, and foreign agriculture

Institution	Address			Research areas
National Research Institute of Agricultural Engineering (NRIAE)	2-1-2 Kannondai, Tsukuba City, Ibaraki Pref. 305, Japan	(0298) 38-7513	(0298) 38-7609	Planning for rural districts, regional resources, hydraulic and structural design
National Research Institute of Aquaculture	422-1 Nakatsuhamaura, Nansei-cho, Watari-gun, Mie Pref. 516-01, Japan	(05996) 6-1830	(05996) 6-1962	Fish genetics, nutrition, and pathology, environmental management
National Research Institute of Brewing (NRIB)	3-7-1 Kagamiyama, Higasihiroshima-shi, Hiroshima 739, Japan	(0824) 20-0800	(0824) 20-0802	Beverage science, environment, microbiology, process engineering
National Research Institute of Far Seas Fisheries (NRIFSF)	7-1 Orido 5-chome, Shimizu City, Shizuoka Pref. 424, Japan	(0543) 36-6000	(0824) 35-9642	North Pacific resources, pelagic fish and oceanic resources, oceanography
National Research Institute of Fire and Disaster (NRIFD)	14-1 Nakahara 3-chome, Mitaka City, Tokyo 181, Japan	(0422) 44-8331	(0824) 42-7719	Evacuation systems, protection, intelligent fire detection systems
National Research Institute of Fisheries Engineering (NRIFE)	Ebidai, Hasaki-machi, Kashima-gun, Ibaraki Pref. 314-04, Japan	(0479) 44-5929	(0479) 44-1875	Aquaculture and fish port engineering, fishing boats and instruments
National Research Institute of Police Science	6 Sanban-cho, Chiyoda-ku, Tokyo 102, Japan	(03) 3261-9986	(03) 3221-1245	Forensic biology and anthropology, voice identification, crime prevention
National Research Institute of Vegetables, Ornamental Plants and Tea (NIVOT)	360 Kusawa, Ano, Mie Pref. 514-23, Japan	(059) 268-1331	(059) 268-1339	Applied plant physiology, protected cultivation, plant protection and soil science, tea agronomy, vegetable breeding
National Research Laboratory of Metrology (NRLM)	1-4 Umezono 1-chome, Tsukuba City, Ibaraki Pref. 305, Japan	(0298) 54-4149	(0298) 54-4135	Quantum, thermophysical, and mechanical metrology, measurement systems
National Space Development Agency of Japan (NASDA)	World Trade Center Building 4-1, Hamamatsu-cho 2-chome, Minato-ku, Tokyo 105-60, Japan	(03) 3438-6111	(03) 5402-6513	Web: www.nasda.go.jp Satellites and space transportation systems, R&D, experiments, launching
NHK Science & Technical Research Laboratories	1-10-11 Kinuta, Setagaya-ku, Tokyo 157, Japan	(03) 5494-2403	(03) 5494-2418	New broadcasting media (DBS, HDTV, ISDB)

* From outside Japan, dial the international access code ("011" in the U.S.) + country code (81) + area code (omit "0"), then rest of number.

Source: Constructed from the information contained in National Laboratories and Public Research Organizations in Japan, published by JST.

National Laboratories and Research Organizations in Japan	Address	Phone Number*	Fax Number*	Typical Research Activities and Web Address (If Available)
Nippon Institute for Biological Science (NIBS)	2221-1 Shinmachi, Ome, Tokyo 198, Japan	(0428) 33-1001	(0428) 31-6166	Viruses pathogenic for mammals, preventive measures
Okayama Ceramics Research Foundation	1406-18 Nishi Katakami, Bizen, Okayama Pref. 705, Japan	(0869) 64-0505	(0869) 63-0227	High-temperature structural materials, welding, composites
Osaka Bioscience Institute (OBI)	6-2-4, Furuedai, Suita-shi, Osaka 565, Japan	(06) 872-4812	(06) 872-4818	Molecular and cell biology, neuroscience
Osaka National Research Institute, AIST	8-31, Midorigaoka 1-chome, Ikeda City, Osaka 563, Japan		(0727) 51-9621	Energy conversion, energy and the environment, optical materials, organic materials, material physics, new materials, interdisciplinary research
Research Planning Office:		(0727) 51-9681		
International Research Cooperation Section:		(0727) 51-9862		
Osaka Prefectural Institute of Public Health	3-69 Nakamichi 1-chome, Higashinari-ku, Osaka 537, Japan	(06) 972-1321	(06) 972-2393	Web: www.iph.pref.osaka.jp Public health, food hygiene, pharmaceuticals
Overseas Coastal Area Development Institute of Japan (OCDI)	Kazan Bldg., 3-2-4 Kasumigaseki, Chiyoda-ku, Tokyo 100, Japan	(03) 3580-3271/3	(03) 3580-3657	Consulting, research, training and seminars
Port and Harbor Research Institute (PHRI)	3-1-1 Nagase, Yokosuka City, Kanagawa Pref. 239, Japan	(0468) 44-5003	(0468) 42-9265	Hydraulic, geotechnical and structural engineering, marine environment
Power Reactor and Nuclear Fuel Development Corporation (PNC)	9-13, 1-chome, Akasaka Minato-ku, Tokyo 107, Japan	(03) 3586-3311	(03) 3583-6386	Fuel cycle technologies, advanced thermal reactors, fast breeder reactors
Public Works Research Institute (PWRI)	1 Asahi, Tsukuba City, Ibaraki Pref. 305, Japan	(0298) 64-2211	(0298) 64-2840	Environment, rivers, water quality control, dams, roads, bridges
Radiation Effects Research Foundation (RERF)	5-2 Hijiyama Park, Minami-ku, Hiroshima 732, Japan	(082) 261-3131	(082) 263-7279	Epidemiological, clinical, and laboratory research on medical effects

Institute	Address			Research
Railway Technical Research Institute (RTRI)	Shinjuku Office: STEC Joho Bldg. 27F, 1-24-1 Nishishinjuku Shinjuku-ku, Tokyo 160, Japan Kunitachi Institute: 2-8-38 Hikari-cho, Kokubunji-shi, Tokyo 185, Japan	(03) 5322-3557 (0425) 73-7258	(03) 5322-3575 (0425) 73-7356	Testing and research on the magnetic levitation system, shinkansen, narrow gauge lines, and basic technologies; surveys and investigation on superconducting; railway technology standardization; information services
Remote Sensing Technology Center of Japan (RESTEC)	Roppongi First Building, 1-9-9 Roppongi, Minato-ku, Tokyo 106, Japan	(03) 5561-9771-7	(03) 5561-9540-2	Sensor performance evaluation, remote sensing data application
Research Institute for Brain and Blood Vessels–Akita	6-10 Senshu-kubota-machi, Akita City, Akita Pref. 010, Japan	(0188) 33-0115	(0188) 33-2104	Pathophysiology of cerebrovascular and heart diseases
Sagami Chemical Research Center (SCRC)	Nishi-Ohnuma 4-4-1, Sagamihara, Kanagawa 229, Japan	(0427) 42-4791	(0427) 49-7631	Pharmaceutical and agricultural chemicals, polymer materials
Saitama Cancer Center Research Institute	818 Komuro, Ina, Kitaadachi-gun, Saitama Pref. 362, Japan	(048) 722-1111	(048) 722-1739	Discovery of causative gene of acute myeloid leukemia, cancer preventive properties of green tea
Seikai National Fisheries Research Institute	49 Kokubu-machi, Nagasaki City, Nagasaki Pref. 850, Japan	(0958) 22-8158	(0958) 21-4494	Fishery biology, population dynamics, biomass estimation
Shikoku National Agricultural Experiment Station	1-3-1 Senyu-cho, Zentsuji City, Kagawa Pref. 765, Japan	(0877) 62-0800	(0877) 63-1683	Labor-saving horticultural systems, crop improvement
Shikoku National Industrial Research Institute	2217-14 Hayashi-cho, Takamatsu City, Kagawa Pref. 761-03, Japan	(0878) 69-3511	(0878) 69-3553	Development of marine resources, absorbents for rare metals
Ship Research Institute (SRI)	38-1, 6-chome, Shinkawa, Mitaka, Tokyo 181, Japan	(0422) 41-3007	(0422) 41-3247	High-quality ships, ocean space utilization, surface effect ships

* From outside Japan, dial the international access code ("011" in the U.S.) + country code (81) + area code (omit "0"), then rest of number.

Source: Constructed from the information contained in National Laboratories and Public Research Organizations in Japan, published by JST.

National Laboratories and Research Organizations in Japan	Address	Phone Number*	Fax Number*	Typical Research Activities and Web Address (If Available)
Superconductivity Research Laboratory (SRL) /International Superconductivity Technology Center (ISTEC)	SRL: 10-13 Shinonome 1-chome, Koto-ku, Tokyo 135, Japan ISTEC: Eishin-Kaihatsu Bldg., 34-3, Shinbashi 5-chome, Minato-ku, Tokyo 105, Japan	(03) 3536-5703 (03) 3431-4002	(03) 3536-5714 (03) 3431-4044	Fundamental properties on high-tech super-conductors, new oxide superconductors, mechanism of high-tech superconductivity, thin-film processing, basic concepts of devices, chemical processing
Technology Center of Nagasaki (TCN)	2-1303-8 Ikeda, Omura City, Nagasaki Pref. 856, Japan	(0957) 52-1133	(0957) 52-1136	Physical and chemical vapor deposition, Omura Bay regeneration
The Institute of Physical and Chemical Research (RIKEN)	Hirosawa 2-1, Wako City, Saitama Pref. 351-01, Japan Tsukuba Life Science Center: Koyadai 3-1-1, Tsukuba City, Ibaraki Pref. 305, Japan	(048) 462-1111 (0298) 36-9009	(048) 462-1554 (0298) 36-2616	Cosmic radiation, radiation, cyclotron, linear accelerator, atomic physics, biophysics, muon science, magnetic materials, information science, and many others
The National Institute of Health and Nutrition	1-23-1 Toyama, Shinjuku-ku, Tokyo 162, Japan	(03) 3203-5721	(03) 3202-3278	Cardiovascular disease, hypertension, diabetes, gout, exercise
The Research Institute of Tuberculosis–Japan Anti-Tuberculosis Association (JATA)	3-1-24 Matsuyama, Kiyose-shi, Tokyo 204, Japan	(0424) 93-5711	(0424) 92-4600	Interpretation of chest X-ray films, pathogenesis, healing mechanisms of tuberculosis lesions, BCG vaccine
The Tokyo Metropolitan Institute of Medical Science	18-22 Honkomagome 3-chome, Bunkyo-ku, Tokyo 113, Japan	(03) 3823-2101	(03) 3823-2965	Web: www.rinshoken.or.jp Molecular biology, physiology, cancer
The Tokyo Metropolitan Research Laboratory of Public Health	3-24-1 Hyakunincho, Shinjuku-ku, Tokyo, Japan	(03) 3363-3231	(03) 3368-4060	Research on public health information, prevention of infections pharmaceuticals
Tohoku National Agricultural Experiment Station	4 Akahira, Shimo-kuriyagawa, Morioka City, Iwate Pref. 020-01, Japan	(019) 643-3433	(019) 641-7794	Rice and other crop farming systems, cattle production

Tohoku National Fisheries Research Institute	27-5 Shinhama-cho, 3-chome, Shiogama City, Miyagi Pref. 985, Japan	(022) 365-1191	(022) 367-1250	Fishery oceanography, resources management and enhancement
Tohoku National Industrial Research Institute	4-2-1 Nigatake, Miyagino-ku, Sendai 983, Japan	(022) 237-5211, or (022) 237-0936 (dial in)	(022) 236-6839	Metals and materials for geothermal power generation, reinforced materials
Tokyo Metropolitan Institute for Neuroscience	2-6 Musashidai, Fuchu City, Tokyo 183, Japan	(0423) 25-3881	(0423) 21-7876	Intractable neurological diseases, development of nervous system
Tokyo Metropolitan Institute of Gerontology	35-2 Sakaecho, Itabashi-ku, Tokyo 173, Japan	(03) 3964-3241	(03) 3579-4776	Web: www.tmig.or.jp Age-associated dementia, longitudinal interdisciplinary study on aging
Tottori Mycological Institute (TMI)	Kokoge-211, Tottori City, Tottori Pref. 689-11, Japan	(0857) 51-8111	(0857) 53-1986	Taxonomy, physiology, and ecology of fungi, biotechnological study
Traffic Safety and Nuisance Research Institute (TSNRI)	6-38-1 Shinkawa, Mitaka City, Tokyo 181, Japan	(0422) 41-3207 (planning office)	(0422) 41-3233	Safety of traffic systems, railways, ropeways and air transport; automotive technology

* From outside Japan, dial the international access code ("011" in the U.S.) + country code (81) + area code (omit "0"), then rest of number.

Source: Constructed from the information contained in National Laboratories and Public Research Organizations in Japan, published by JST.

Appendix 3a
Selected Research-Oriented National Universities in Japan

Japanese National Universities and Their Attached Institutes and Laboratories (Science and Technology Only)	Address*	Phone Number[†]	Fax Number[†]
Hokkaido University	Nishi 5-chome, Kita 8-jou, Kita-ku, Sapporo, Hokkaido 060-0808, Japan	(011) 716-2111	
The Institute of Low-Temperature Science	Nishi 8-chome, Kita 19-jou, Kita-ku, Sapporo, Hokkaido 060, Japan	(011) 716-2111	(011) 716-5698
Research Institute for Electronic Science	Nishi 6-chome, Kita 12-jour, Kita-ku, Sapporo, Hokkaido 060, Japan	(011) 716-2111	(011) 758-6098
The Research Institute for Catalysis	Nishi 10-chome, Kita 11-jou, Kita-ku, Sapporo, Hokkaido 060, Japan	(011) 716-2111	(011) 709-4748
Institute of Immunological Sciences	Nishi 7-chome, Kita 15-jou, Kita-ku, Sapporo, Hokkaido 060, Japan	(011) 716-2111	(011) 758-7568
Muroran Institute of Technology	27-1, Mizumoto-cho, Muroran, Hokkaido 050-8585	(0143) 44-4181	(0143) 47-3126
Hirosaki University	1 Bunkyo-cho, Hirosaki, Aomori 036-8560, Japan	(0172) 36-2111	(0172) 37-6594
Tohoku University	2-1-1 Katahira, Aoba-ku, Sendai, Miyagi 980-8577, Japan	(022) 217-5900	(022) 217-5030
The Institute for Materials Research	2-1-1 Katahira, Aoba-ku, Sendai, Miyagi 980, Japan	(022) 217-5900	
The Institute for Advanced Materials Processing	2-1-1 Katahira, Aoba-ku, Sendai, Miyagi 980, Japan	(022) 217-5900	
The Institute of Development, Aging and Cancer	4-1, Seiryo, Aoba-ku, Sendai, Miyagi 980, Japan	(022) 717-7000	
Institute of Genetic Ecology	4-1, Seiryo, Aoba-ku, Sendai, Miyagi 980, Japan	(022) 717-7000	
Research Institute for Scientific Measurements	2-1-1 Katahira, Aoba-ku, Sendai, Miyagi 980, Japan	(022) 217-5900	
The Institute of Fluid Science	2-1-1 Katahira, Aoba-ku, Sendai, Miyagi 980, Japan	(022) 217-5900	(022) 223-2748
Research Institute of Electrical Communication	2-1-1 Katahira, Aoba-ku, Sendai, Miyagi 980, Japan	(022) 217-5900	(022) 224-7889

* The addresses and phone numbers listed for the universities are those of the secretariats of the graduate schools.
[†] From outside Japan, dial the international access code ("011" in the U.S.) + country code (81) + area code (omit "0"), then rest of number.

Japanese National Universities and Their Attached Institutes and Laboratories (Science and Technology Only)	Address*	Phone Number†	Fax Number†
Institute for Chemical Reaction Science	2-1-1 Katahira, Aoba-ku, Sendai, Miyagi 980, Japan	(022) 217-5900	(022) 223-8956
Ibaraki University	2-1-1 Bunkyo, Mito, Ibaraki 310-8512, Japan	(029) 226-1621	(029) 228-8019
The University of Tsukuba	1-1-1 Tenno-dai, Tsukuba, Ibaraki 305-8577, Japan	(0298) 53- 2111	(0298) 53-6019
Gunma University	4-2, Aramaki-cho, Maebashi, Gunma 371-8510, Japan	(027) 220-7111	(027) 220-7012
Institute of Endocrinology	3-39-15 Showa-cho, Maebashi, Gunma 371, Japan	(027) 231-7221	(027) 234-1788
Saitama University	225 Shimo-Okubo, Urawa, Saitama 338-8570, Japan	(048) 852-2111	(048) 852-3677
Chiba University	1-33 Yayoi-cho, Inage-ku, Chiba 263-8522, Japan	(043) 251-1111	(043) 290-2011
The University of Tokyo	7-3-1 Hongo, Bunkyo-ku, Tokyo 113-8654, Japan	(03) 3812-2111	(03) 3816-3913
The Institute of Medical Science	4-6-1 Shiroganedai, Minato-ku, Tokyo 108, Japan	(03) 3443-8111	(03) 3443-3893
Earthquake Research Institute	1-1-1 Yayoi, Bunkyo-ku, Tokyo 113, Japan	(03) 3812-2111	(03) 3816-1159
Institute of Industrial Science	7-22-1 Roppongi, Minato-ku, Tokyo 106, Japan	(03) 3402-6231	
Institute of Molecular and Cellular Biosciences	1-1-1 Yayoi, Bunkyo-ku, Tokyo 113, Japan	(03) 3812-2111	(03) 3815-6805
Institute for Cosmic Ray Research	3-2-1 Midori-cho, Tanashi, Tokyo 188, Japan	(0424) 61-4131	(0424) 68-1438
Institute for Nuclear Study	3-2-1 Midori-cho, Tanashi, Tokyo 188, Japan	(0424) 61-4131	(0424) 64-3212
The Institute for Solid State Physics	7-22-1 Roppongi, Minato-ku, Tokyo 106, Japan	(03) 3478-6811	(03) 3401-5169
Ocean Research Institute	1-15-1 Minamidai, Nakano-ku, Tokyo 164, Japan	(03) 3376-1251	(03) 3375-6716
Tokyo Medical and Dental University	1-5-45 Yushima, Bunkyo-ku, Tokyo 113-0034, Japan	(03) 3813-6111	(03) 5684-0198
Institute for Medical and Dental Engineering	2-3-10 Kanda Surugadai, Chiyoda-ku, Tokyo 101, Japan	(03) 3291-3721	(03) 3291-3727
Medical Research Institute	2-3-10 Kanda Surugadai, Chiyoda-ku, Tokyo 101, Japan	(03) 3294-7311	(03) 3294-7316
Tokyo Institute of Technology	2-12-1 Oh-okayama, Meguro-ku, Tokyo 152-8550, Japan	(03) 3726-1111	(03) 5734-3445

* The addresses and phone numbers listed for the universities are those of the secretariats of the graduate schools.
† From outside Japan, dial the international access code ("011" in the U.S.) + country code (81) + area code (omit "0"), then rest of number.

Japanese National Universities and Their Attached Institutes and Laboratories (Science and Technology Only)	Address*	Phone Number[†]	Fax Number[†]
Research Laboratory of Resources Utilization	4259 Nagatsuda-cho, Midori-ku, Yokohama, Kanagawa 227, Japan	(045) 922-1111	(045) 921-0897
Precision and Intelligence Laboratory	4259 Nagatsuda-cho, Midori-ku, Yokohama, Kanagawa 227, Japan	(045) 922-1111	(045) 921-0898
Materials and Structures Laboratory	4259 Nagatsuda-cho, Midori-ku, Yokohama, Kanagawa 227, Japan	(045) 922-1111	(045) 921-1015
Research Laboratory for Nuclear Reactors	2-12-1 Oh-Okayama, Meguro-ku, Tokyo 152, Japan	(03) 3726-1111	(03) 3729-1879
The University of Electro-Communications	1-5-1 Chofugaoka, Chofu, Tokyo 182-8585, Japan	(0424) 83-2161	(0424) 81-3612
Tokyo University of Agriculture and Technology	3-8-1, Harumi-cho, Fuchu, Tokyo 183-0057, Japan	(042) 367-5504	(042) 367-5553
Yokohama National University	79-1 Tokiwa-dai, Hodogaya-ku, Yokohama 240-8501, Japan	(045) 335-1451	(045) 341-2582
Niigata University	8050 Igarashi Ninomachi, Niigata 950-2181, Japan	(025) 262-6098	(025) 262-6539
Brain Research Institute	757, 1-bancho, Asahimachi-dori, Niigata, Niigata 951, Japan	(025) 223-6161	(025) 225-6458
Nagaoka University of Technology	1603-1, Kami-Tomiokacho, Nagaoka,, Niigata 940-2188, Japan	(0258) 46-6000	(0258) 46-7363
Toyama University	3190 Gofuku, Toyama, Toyama 930-8555, Japan	(0764) 45-6123	
Toyama Medical and Pharmaceutical University	2630 Sugitani, Toyama, Toyama 930-0194, Japan	(0764) 34-2281	(0764) 34-1463
Research Institute for WAKAN-YAKU	2630 Sugitani, Toyama, Toyama 930-0194, Japan	(0764) 34-2281	
Kanazawa University	Kadoma-machi, Kanazawa, Ishikawa 920-1192, Japan	(076) 264-5111	(076) 234-4010
Cancer Research Institute	13-1 Takara-machi, Kanazawa, Ishikawa 920, Japan	(0762) 62-8151	(0762) 22-5831
Shinshu University	3-1-1 Asahi, Matsumoto, Nagano 390-8621, Japan	(0263) 35-4600	(0263) 36-6769
Shizuoka University	836 Otani, Shizuoka, Shizuoka 422-8529, Japan	(054) 237-1111	(054) 237-0089
Research Institute of Electronics	3-5-1 Johoku, Hamamatsu, Shizuoka 432, Japan	(0534) 71-1171	(0534) 74-0630
Gifu University	1-1 Yanagito, Gifu, Gifu 501-1193, Japan	(058) 230-1111	(058) 293-2021

* The addresses and phone numbers listed for the universities are those of the secretariats of the graduate schools.
† From outside Japan, dial the international access code ("011" in the U.S.) + country code (81) + area code (omit "0"), then rest of number.

Japanese National Universities and Their Attached Institutes and Laboratories (Science and Technology Only)	Address*	Phone Number†	Fax Number†
Nagoya University	Shiranui-cho, Senju-ku, Nagoya, Aichi 464-8601, Japan	(052) 781-5111	
The Research Institute of Environmental Medicine	Shiranui-cho, Senju-ku, Nagoya, Aichi 464-8601, Japan	(052) 781-5111	(052) 781-9117
Institute for Hydrospheric-Atmospheric Sciences	13, 3-chome, Honohara, Toyokawa, Aichi 442, Japan	(053) 386-3154	(053) 386-0811
Institute of Plasma Physics	Shiranui-cho, Senju-ku, Nagoya, Aichi 464-8601, Japan	(052) 781-5111	(052-782-7106
Water Research Institute	Shiranui-cho, Senju-ku, Nagoya, Aichi 464-8601, Japan	(052) 781-5111	(052) 781-3998
Nagoya Institute of Technology	Mikidokoro-cho, Showa-ku, Nagoya, Aichi 466-8555, Japan	(052) 732-2111	(052) 735-5009
Toyohashi University of Technology	1-1 Hibarigaoka, Tenpaku-cho, Toyohashi, Aichi 441-8580, Japan	(0532) 47-0111	
Mie University	1515 Kamihama-cho, Tsu, Mie 514-8507, Japan	(059) 232-1211	(059) 231-9000
Kyoto University	Yoshida Honmachi, Sakyo-ku, Kyoto, Kyoto 606-8501, Japan	(075) 753-7531	(075) 753-2092
The Institute for Chemical Research	Gokasho, Uji, Kyoto 611, Japan	(0774) 32-3111	(0774) 32-1247
Institute for Frontier Medical Sciences	53 Shogoin Kawahara-cho, Sakyo-ku, Kyoto, Kyoto 606, Japan	(075) 751-3802	(075) 752-9017
Institute of Advanced Energy	Gokasho, Uji, Kyoto 611, Japan	(0774) 32-3111	(0774) 33-3234
Wood Research Institute	Gokasho, Uji, Kyoto 611, Japan	(0774) 32-3111	(0774) 33-3049
Research Institute for Food Science	Gokasho, Uji, Kyoto 611, Japan	(0774) 32-3111	(0774) 33-3004
Disaster Prevention Research Institute	Gokasho, Uji, Kyoto 611, Japan	(0774) 32-3111	(0774) 32-4115
Yukawa Institute for Theoretical Physics	Kitashirakawa Oiwake-cho, Sakyo-ku, Kyoto, Kyoto 606, Japan	(075) 753-7003	(075) 753-7010
Institute for Virus Research	Shogoin Kawahara-cho, Sakyo-ku, Kyoto, Kyoto 606, Japan	(075) 751-3111	(075) 761-5626
Research Institute for Mathematical Sciences	Kitashirakawa Oiwake-cho, Sakyo-ku, Kyoto, Kyoto 606, Japan	(075) 753-7202	(075) 753-7272
Research Reactor Institute	Oaza Noda, Kumatori-cho, Sennangun, Osaka 590-04, Japan	(0724) 52-0901	(0724) 53-1043
Primate Research Institute	Kanbayashi, Inuyama, Aichi 484, Japan	(0568) 61-2891	(0568) 62-2428

* The addresses and phone numbers listed for the universities are those of the secretariats of the graduate schools.
† From outside Japan, dial the international access code ("011" in the U.S.) + country code (81) + area code (omit "0"), then rest of number.

Japanese National Universities and Their Attached Institutes and Laboratories (Science and Technology Only)	Address*	Phone Number[†]	Fax Number[†]
Kyoto Institute of Technology	Matsugasaki, Sakyo-ku, Kyoto, Kyoto 606-8585, Japan	(075) 724-7111	(075) 724-7010
Osaka University	1-1 Yamadaoka, Suita, Osaka 565-0871, Japan	(06) 877-5111	(06) 879-7008
Research Institute for Microbial Diseases	3-1 Yamadaoka, Suita, Osaka 565, Japan	(06) 877-5121	(06) 876-2678
The Institute of Scientific and Industrial Research	8-1 Mihogaoka, Ibaraki, Osaka 567, Japan	(06) 877-5111	(06) 877-4977
Institute for Protein Research	3-2 Yamadaoka, Suita, Osaka 565, Japan	(06) 877-5111	(06) 876-2533
Joining and Welding Research Institute	8-1 Mihogaoka, Ibaraki, Osaka 567, Japan	(06) 877-5111	(06) 877-4594
Kobe University	1-1 Rokkodai-cho, Nada-ku, Kobe, Hyogo 657-8501, Japan	(078) 881-1212	(078) 803-0055
Nara Institute of Science and Technology	8916-5, Takayama-cho, Ikoma, Nara 630-0101, Japan	(0743) 72-5111	
Tottori University	4-101, Koyama-cho Minami, Tottori, Tottori 680-0945, Japan	(0857) 28-0321	(0857) 31-5018
Okayama University	1-1-1, Tsushima Naka, Okayama, Okayama 700-8530, Japan	(086) 252-1111	(086) 251-7019
Research Institute for Bioresources	20-1, 2-chome, Chuo, Kurashiki, Okayama 710, Japan	(0864) 24-1661	(0864) 21-0699
Hiroshima University	1-3-2 Kagamiyama, Higashihiroshima, Hiroshima, 739-8511, Japan	(0824) 22-7111	
Fisheries Laboratory	1294 Takehara-cho, Takehara, Hiroshima 725, Japan	(0846) 22-2362	(0846) 23-0038
Research Institute for Nuclear Medicine and Biology	1-2-3 Kasumi, Minami-ku, Hiroshima, Hiroshima 734-8553, Japan	(082) 257-5555	(082) 255-8339
Yamaguchi University	1677-1, Yoshida, Yamaguchi, Yamaguchi 753-8511, Japan	(0839) 33-5000	
Kagawa University	1-1, Saiwai-cho, Takamatsu, Kagawa 760-8512, Japan	(087) 832-1000	
Ehime University	10-13, Dougo-Himata, Matsuyama, Ehime 790-8577	(089) 927-9000	
Kyushu University	6-10-1, Hakozaki, Higashi-ku, Fukuoka, Fukuoka 812-8581, Japan	(092) 642-2111	
Medical Institute of Bioregulation	3-1-1 Umade, Higashi-ku, Fukuoka, Fukuoka 812, Japan	(092) 641-1151	(092) 641-1315
Institute of Health Science	3-1-1 Umade, Higashi-ku, Fukuoka, Fukuoka 812, Japan	(092) 641-1151	
Institute of Advanced Material Study	6-1 Kasuga Koen, Kasuga, Fukuoka 816, Japan	(092) 573-9611	

* The addresses and phone numbers listed for the universities are those of the secretariats of the graduate schools.
[†] From outside Japan, dial the international access code ("011" in the U.S.) + country code (81) + area code (omit "0"), then rest of number.

Japanese National Universities and Their Attached Institutes and Laboratories (Science and Technology Only)	Address*	Phone Number†	Fax Number†
Research Institute for Applied Mechanics	6-1 Kasuga Koen, Kasuga, Fukuoka 816, Japan	(092) 573-9611	(092) 573-6899
Nagasaki University	1-14, Bunkyo-cho, Nagasaki, Nagasaki 852-8521, Japan	(095) 847-1111	(095) 844-2349
Institute for Tropical Medicine	12-4 Sakamoto-cho, Nagasaki, Nagasaki 852-8521, Japan	(0958) 47-2111	(0958) 47-6607
Kumamoto University	2-39-1, Kurokami, Kumamoto, Kumamoto 860-8555, Japan	(096) 344-2111	(096) 342-3110
Kagoshima University	1-21-24, Koorimoto, Kagoshima, Kagoshima 890-0065, Japan	(099) 285-7111	(099) 285-7034
Ryukyu University	1, Chihara, Nishihara-machi, Nakagami-gun, Okinawa 903-0213, Japan	(098) 895-2221	(098) 895-4586
Common Research Organizations for Japanese National Universities			
National Laboratory for High Energy Physics	1-1 Ooho, Tsukuba, Ibaragi 305, Japan	(0298) 64-1171	(0298) 64-2397
National Institute of Polar Research	1-9-10 Kaga, Itabashi-ku, Yokyo 173, Japan	(03) 3962-4711	(03) 3962-2529
The Institute of Space and Astronautical Science	3-1-1 Yunodai, Sagamihara, Kanagawa 229, Japan	(0427) 51-3911	(0427) 59-4255
National Institute of Genetics	1111, Tanida, Mishima, Shizuoka 411, Japan	(0559) 81-6707	(0559) 81-6715
The Institute of Statistical Mathematics	4-6-7, Minami Azabu, Minato-ku, Tokyo 106, Japan	(03) 3446-1501	(03) 3446-1695
National Institute for Fusion Science	322-6, Shoimoishi-cho, Doki, Gifu 509-5292, Japan	(0572) 58-2222	
Okazaki National Research Institutes	38 Aza Saigonaka, Myodaiji-cho, Okazaki, Aichi 444-8585, Japan	(0564) 55-7000	(0564) 52-4889
Institute for Molecular Science	38 Aza Saigonaka, Myodaiji-cho, Okazaki, Aichi 444-8585, Japan	(0564) 55-7000	(0564) 55-7119
National Institute for Basic Biology	38 Aza Saigonaka, Myodaiji-cho, Okazaki, Aichi 444-8585, Japan	(0564) 55-7000	(0564) 55-7119
National Institute for Physiological Sciences	38 Aza Saigonaka, Myodaiji-cho, Okazaki, Aichi 444-8585, Japan	(0564) 55-7000	(0564) 55-7119
National Center for Science Information System	3-29-1 Otsuka, Bunkyo-ku, Tokyo 112-8640, Japan	(03) 3942-2351	(03) 3942-6900
National Museum of Ethnology	10-1 Senri Banpaku Koen, Suita, Osaka 565-8511, Japan	(06) 876-2151	(06) 875-0401
National Museum of Japanese History	117 Jonai-cho, Sakura, Chiba 285-0017, Japan	(043) 486-0123	(043) 486-4209
National Astronomical Observatory	2-21-1 Mitaka, Tokyo 181, Japan	(0422) 41-3600	(0422) 34-3690

* The addresses and phone numbers listed for the universities are those of the secretariats of the graduate schools.
† From outside Japan, dial the international access code ("011" in the U.S.) + country code (81) + area code (omit "0"), then rest of number.

Appendix 3b
Selected Public Universities Established by Local Governments

Japanese Public Universities and Their Attached Institutes	Address*	Phone Number†
Sapporo Medical College	Nishi 17-291, Minami 1-jou, Chuo-ku, Sapporo, Hokkaido 060-0061, Japan	(011) 611-2111
Gan Kenkyusho (Cancer Research Institute)	Nishi 17-291, Minami 1-jou, Chuo-ku, Sapporo, Hokkaido 060-0061, Japan	(011) 611-2111
Fukushima Medical College	1 Hikarigaoka, Fukushima, Fukushima 960-1295, Japan	(024) 548-2111
Seitai Jouhou Dentatsu Ken-kyusho (Institute for Biological Information Transmission)	1 Hikarigaoka, Fukushima, Fukushima 960-1295, Japan	(024) 548-2111
Radio Isotope Research Facility	1 Hikarigaoka, Fukushima, Fukushima 960-1295, Japan	(024) 548-2111
Tokyo Metropolitan University	Minami Osawa, Hachioji, Tokyo 192-0397, Japan	(0426) 77-1111
Tokyo Metropolitan Institute of Technology	6-6 Asahigaoka, Hino, Tokyo 191-0065, Japan	(042) 585-8600
Yokohama City University	22-2 Seto, Kanazawa-ku, Yokohama, Kanagawa 236-0027, Japan	(045) 787-2311
Kihara Seibutsugaku Kenkyusho (Kihara Biology Institute)		
Nagoya City University	1 Kawazumi, Mizuho-cho, Mizuho-ku, Nagoya, Aichi 467-0001, Japan	(052) 851-5511
Osaka Prefecture University	Gakuen-machi, Sakai, Osaka 599-8531, Japan	(0722) 52-1161
Sentankagaku Kenkyusho (Frontier Science Institute)	Gakuen-machi, Sakai, Osaka 599-8531, Japan	(0722) 52-1161
Osaka City University	3-3-138, Sugimoto, Sumiyoshi-ku, Osaka 558-8585, Japan	(06) 605-2011
Himeji Institute of Technology	2167 Shosha, Himeji, Hyogo 671-2201, Japan	(0792) 66-1661
Wakayama Medical College	27, Kyuban-cho, Wakayama, Wakayama 640-8155, Japan	(0734) 31-2151
Ohyou Igaku Kenkyusho (Institute for Applied Medicine)	27, Kyuban-cho, Wakayama, Wakayama 640-8155, Japan	(0734) 31-2151

* The addresses and phone numbers listed for the universities are those of the secretariats of the graduate schools.
† From outside Japan, dial the international access code ("011" in the U.S.) + country code (81) + area code (omit "0"), then rest of number.
Note: Names in parentheses are unofficial English translations by the editor.

Appendix 3c

Selected Research Oriented Private Universities in Japan

Japanese Private Universities and Their Attached Institutes and Laboratories (Science and Technology Only)	Address*	Phone Number[†]	Fax Number[†]
Dohto University	7-1, Ochiishi-cho, Monbetsu, Hokkaido 094-8582, Japan	(01582) 4-8101	(01582) 4-8101
Marine Biology Research Center	7-chome, Ochiishi-cho, Monbetsu, Hokkaido 094-8582, Japan	(01582) 3-4707	
Hokkai Gakuen University	1-40, 4-chome, Asahi-cho, Toyohira-ku, Sapporo, Hokkaido 062-8605, Japan	(011) 841-1161	(011) 824-3141
Development Research Institute	1-40, 4-chome, Asahi-cho, Toyohira-ku, Sapporo, Hokkaido 062-8605, Japan	(011) 841-1161	
Tohoku Institute of Technology	35-1, Kasumi-cho, Yagiyama, Tashiro-ku, Sendai, Miyagi 982-8577	(022) 229-1151	(022) 228-2781
Informatics Laboratory	35-1 Yagiyama Kasumi-cho, Tashiro-ku, Sendai, Miyagi 982, Japan	(022) 229-1151	(022) 229-0005
Tohoku College of Pharmacy	4-4-1 Komatsujima, Aoba-ku, Sendai, Miyagi 981-8558, Japan	(022) 234-4181	(022) 275-2013
Cancer Research Institute	4-4-1 Komatsujima, Aoba-ku, Sendai, Miyagi 981-8558, Japan	(022) 234-4181	(022) 275-2013
Josai University	1-1, Keyaki-dai, Sakado, Saitama 350-0295, Japan	(0492) 86-2233	
Center for Computer Science	1-1 Keyakidai, Sakado, Saitama 350-0295, Japan	(0492) 86-2233	(0492) 85-7167
Chiba Institute of Technology	2-17-1 Tsudanuma, Narashino, Chiba 275-0016, Japan	(047) 475-2111	(047) 478-0259
Electronic Computing Center	2-17-1 Tsudanuma, Narashino, Chiba 275-0016, Japan	(047) 478-0570	
Aoyama Gakuin University	4-4-25 Shibuya, Shibuya-ku, Tokyo 150-8366, Japan	(03) 3409-8111	
Information Science Research Center	4-4-25 Shibuya, Shibuya-ku, Tokyo 150-8366, Japan	(03) 3409-8111	(03) 3498-4870

* The addresses and phone numbers listed for the universities are those of the secretariats of the graduate schools.

[†] From outside Japan, dial the international access code ("011" in the U.S.) + country code (81) + area code (omit "0"), then rest of number.

Japanese Private Universities and Their Attached Institutes and Laboratories (Science and Technology Only)	Address*	Phone Number[†]	Fax Number[†]
Gakushuin University	1-5-1 Mejiro, Tohima-ku, Tokyo 171-8588, Japan	(03) 3986-0221	
Computer Center	1-5-1 Mejiro, Tohima-ku, Tokyo 171-8588, Japan	(03) 3986-0221	(03) 3992-1018
Keio University	2-15-45 Mita, Minato-ku, Tokyo 108-8345, Japan	(03) 3453-4511	(03) 3769-1564
Pharmaceutical Institute, School of Medicine	35 Shinano-machi, Shinjuku-ku, Tokyo 160, Japan	(03) 3353-1211	
Komazawa University	1-23-1 Komazawa, Setagaya-ku, Tokyo 154-8525, Japan	(03) 3418-9111	
Institute for Applied Geography	1-23-1 Komazawa, Setagaya-ku, Tokyo 154-8525, Japan	(03) 3418-9553	
Toyo University	5-28-20, Hakusan, Bunkyo-ku, Tokyo 112-8606, Japan	(03) 3945-7224	(03) 3945-7221
Takushoku University	3-4-14, Kohinata, Bunkyo-ku, Tokyo 112-8585, Japan	(03) 3947-2261	
Shibaura Institute of Technology	3-9-14 Shibaura, Minato-ku, Tokyo 108-8548, Japan	(03) 3452-3201	(03) 5476-3181
Research Laboratory of Engineering	3-9-14 Shibaura, Minato-ku, Tokyo 108-8548, Japan	(03) 5476-3110	(03) 5476-3176
Sophia University	7-1 Kioicho, Chiyoda-ku, Tokyo 102-8554, Japan	(03) 3238-3111	
Life Science Institute	7-1 Kioicho, Chiyoda-ku, Tokyo 102-8554, Japan	(03) 3238-3488	(03) 3238-3885
Teikyo University	2-11-1 Kaga, Itabashi-ku, Tokyo 173-8605, Japan	(03) 3964-1211	(03) 3964-8415
Biotechnology Research Center	Sagamiko-cho, Tsukui-gun, Kanagawa 199-01	(04268) 5-1121	(04268) 5-1669
Tokai University	2-28-4 Tomigaya, Shibuya-ku, Tokyo 151-0063, Japan	(03) 3467-2211	
General Research Organization	2-28-4 Tomigaya, Shibuya-ku, Tokyo 151-0063, Japan	(03) 3467-2211	(03) 3460-4515
The Institute of Industrial Science	1117 Kita Kinmoku, Hiratsuka, Kanagawa 259-12	(0463) 58-1211	(0463) 58-1812
Institute of Oceanic Research and Development	3-20-1 Orito, Shimizu, Shizuoka 424, Japan	(0543) 34-0411	(0543) 34-5095
Research and Development Institute	2-28-4 Tomigaya, Shibuya-ku, Tokyo 151-0063, Japan	(03) 3467-2211	(03) 3458-1203
Medical Research Institute	Bouseidai, Isehara, Kanagawa 259-11, Japan	(0463) 93-1121	(0463) 93-1121, ext. 2065
Research and Information Center	2-28-4 Tomigaya, Shibuya-ku, Tokyo 151-0063, Japan	(03) 3481-0611	(03) 3481-0610

Institution	Address	Phone	Phone
Space Information Center	Sugido, Masushiro-cho, Masushiro-gun Kumamoto 861-23, Japan	(096) 286-2929	(096) 286-8801
Hokkaido Tokai University	224 Chuwa, Kami-cho, Asahikawa, Hokkaido 070-8601, Japan	(0166) 61-5111	
Research Institute of Life in Northern Japan	224 Chuwa, Kami-cho, Asahikawa, Hokkaido 070-8601, Japan	(0166) 61-5111	
Kyushu Tokai University	223 Watashika, Ooe-cho, Kumamoto, Kumamoto 862-8652, Japan	(096) 382-1141	
Institute of Industrial Science and Technical Research	223 Watashika, Ooe-cho, Kumamoto, Kumamoto 862-8652, Japan	(096) 382-1141	(096) 381-7956
Technical Center of Knowledge and Information Developments	223 Watashika, Ooe-cho, Kumamoto, Kumamoto 862-8652, Japan	(096) 382-1141	(096) 381-7956
Agricultural Research Institute	Kawahi, Choyo-mura, Aso-gun, Kumamoto 869-14, Japan	(09676) 7-0611	(09676) 7-2053
Junior College of Tokai University	2-3-23 Takanawa, Minato-ku, Tokyo 108-8619, Japan	(03) 3441-1171	
Research Institute of Information and Communications Technology	2-3-23 Takanawa, Minato-ku, Tokyo 108-8619, Japan	(03) 3441-1171	(03) 3447-6005
Research Institute of Domestic Science	101 Miyamae-cho, Shizuoka, Shizuoka 420, Japan	(0542) 61-6321	(0542) 62-9911
Kogakuin University	1-24-2, Nishi-Shinjuku, Shinjuku-ku, Tokyo 163-8677, Japan	(03) 3342-1211	(03) 3342-5304
Tokyo Women's Medical College	8-1, Kawata-cho, Shinjuku-ku, Tokyo 162-8666, Japan	(03) 3353-8111	
The Heart Institute of Japan Hospital	8-1, Kawata-cho, Shinjuku-ku, Tokyo 162-8666, Japan	(03) 3353-8111	(03) 3353-6793
Institute of Gastroenterology	8-1, Kawata-cho, Shinjuku-ku, Tokyo 162-8666, Japan	(03) 3353-8111	
Neurological Institute	8-1, Kawata-cho, Shinjuku-ku, Tokyo 162-8666, Japan	(03) 3353-8111	
Kidney Center	8-1, Kawata-cho, Shinjuku-ku, Tokyo 162-8666, Japan	(03) 3353-8111	
Diabetes Center	8-1, Kawata-cho, Shinjuku-ku, Tokyo 162-8666, Japan	(03) 3353-8111	
Institute of Clinical Endocrinology	8-1, Kawata-cho, Shinjuku-ku, Tokyo 162-8666, Japan	(03) 3353-8111	
Maternal and Perinatal Center	8-1, Kawata-cho, Shinjuku-ku, Tokyo 162-8666, Japan	(03) 3353-8111	(03) 3355-3050
The Institute of Geriatrics	2-15-1 Shibuya, Shibuya-ku, Tokyo 150, Japan	(03) 3499-1911	
Institute of Rheumatology	Shinjuku NS Bldg. 2-4-1, Nishi Shinjuku, Shinjuku-ku, Tokyo 162, Japan	(03) 3348-0988	(03) 3346-2380

* The addresses and phone numbers listed for the universities are those of the secretariats of the graduate schools.
† From outside Japan, dial the international access code ("011" in the U.S.) + country code (81) + area code (omit "0"), then rest of number.

Japanese Private Universities and Their Attached Institutes and Laboratories (Science and Technology Only)	Address*	Phone Number†	Fax Number†
Medical Research Institute	8-1, Kawata-cho, Shinjuku-ku, Tokyo 162-8666, Japan	(03) 3353-8111	(03) 3353-6793
Institute of Biomedical Engineering	8-1, Kawata-cho, Shinjuku-ku, Tokyo 162-8666, Japan	(03) 3353-8111	(03) 3353-6046
Institute of Laboratory Animals	8-1, Kawata-cho, Shinjuku-ku, Tokyo 162-8666, Japan	(03) 3353-8111	
Tokyo Denki University	2-2 Kanda Nishiki-cho, Chiyoda-ku, Tokyo 101-8457, Japan	(03) 5280-3311	(03) 5280-3599
Center for Research	2-2 Kanda Nishiki-cho, Chiyoda-ku, Tokyo 101-8457, Japan	(03) 5280-3311	
Tokyo University of Agriculture	1-1-1 Sakuraoka, Setagaya-ku, Tokyo 156-8502, Japan	(03) 5477-2220	
Institute of Animal Serology	1-1-1 Sakuraoka, Setagaya-ku, Tokyo 156-8502, Japan	(03) 5420-2131, ext. 619	
NODAI Research Institute	1-1-1 Sakuraoka, Setagaya-ku, Tokyo 156-8502, Japan	(03) 5426-7458	(03) 5420-5131
Science University of Tokyo	1-3 Kagurazaka, Shinjuku-ku, Tokyo 162-8601, Japan	(03) 3260-4271	
Research Institute for Science and Technology	Yamasaki, Noda, Chiba 278, Japan	(0471) 24-1501	(0471) 24-2150
Nihon University	4-8-24, Kudan Minami, Chiyoda-ku, Tokyo 102-8275, Japan	(03) 5275-8110	
University Research Center	4-8-24, Kudan Minami, Chiyoda-ku, Tokyo 102-8275, Japan	(03) 5262-2271	(03) 5265-8968
Research Institute of Science and Technology	1-8 Kanda Surugadai, Chiyoda-ku, Tokyo 101, Japan	(03) 3293-3251	(03) 3293-5829
Atomic Energy Research Institute	1-8 Kanda Surugadai, Chiyoda-ku, Tokyo 101, Japan	(03) 3293-3251	(03) 3293-8269
Dental Research Center	1-8-13 Kanda Surugadai, Chiyoda-ku, Tokyo 101, Japan	(03) 3293-5711	(03) 3233-0159
Nippon Medical School	1-1-5, Sendagi, Bunkyo-ku, Tokyo 113-8602, Japan	(03) 3822-2131	
Institute of Gerontology	1-10-19 Ueno Sakuragi, Taito-ku, Tokyo 110, Japan	(03) 3821-4004	(03) 5685-3079
The Research Institute of Vaccine, Therapy for Tumors, and Infectious Diseases	1-1-5 Sendagi, Bunkyo-ku, Tokyo 113, Japan	(03) 3822-2131	(03) 824-6400
Nippon Dental University	1-9-20, Fujimi, Chiyoda-ku, Tokyo 102-8159, Japan	(03) 3261-8311	
Dental Research Institute	1-9-20, Fujimi, Chiyoda-ku, Tokyo 102, Japan	(03) 3261-8311	
Hosei University	2-17-1, Fujimi, Chiyoda-ku, Tokyo 102-0071, Japan	(03) 3264-9308	

Computer Center	3-7-2 Kajino-cho, Koganei, Tokyo 184, Japan	(0423) 81-5341	(0423) 85-9569
Research Center of Ion Beam Technology	3-7-2 Kajino-cho, Koganei, Tokyo 184, Japan	(0423) 81-5341	(0423) 85-9569
Chuo University	742-1, Higashi Nakano, Hachioji, Tokyo 192-0393, Japan	(0426) 74-2111	
Hoshi University	2-4-41 Ebara, Shinagawa-ku, Tokyo 142-8501, Japan	(03) 3786-1011	
Institute of Medicinal Chemistry	2-4-41 Ebara, Shinagawa-ku, Tokyo 142-8501, Japan	(03) 3786-1011	(03) 3787-0036
The Advanced Research Institute for Science and Engineering	1-26-1 Toyotama-kami, Nerima-ku, Tokyo 176-8534, Japan	(03) 5984-3713	
Nezu Chemical Institute	1-26-1 Toyotama-kami, Nerima-ku, Tokyo 176-8534, Japan	(03) 5991-1198	
Musashi Institute of Technology	1-28-1 Tamatsutsumi, Setagaya-ku, Tokyo 158-8557, Japan	(03) 3703-3111	
Atomic Energy Research Laboratory	971 Ozenji, Asoh-ku, Kawasaki, Kanagawa 215, Japan	(044) 966-6131	(044) 955-6071
Meiji University	1-1 Kanda Surugadai, Chiyoda-ku, Tokyo 101-8301, Japan	(03) 3296-4545	
Institute of Science and Technology	1-1-1 Hibashimital, tama-ku, Kawasaki, Kanagawa 214, Japan	(044) 911-8181	
Rikkyo University	3-34-1 Nishi-Ikebukuro, Toshima-ku, Tokyo 171-8501, Japan	(03) 3985-2231	
Institute for Atomic Energy	2-5-1 Nagasaka, Yokosuka, Kanagawa 240-01, Japan	(0468) 56-3131	(0468) 56-7576
Waseda University	1-104 Totsuka-cho, Shinjuku-ku, Tokyo 169-0071, Japan	(03) 3203-4141	
Science and Engineering Research Laboratory	17 Kikui-cho, Shinjuku-ku, Tokyo 162, Japan	(03) 3203-4141, ext. 75-2121	(03) 3203-4141, ext. 75-4480
Kagami Memorial Laboratory for Materials Science and Technology	2-8-26 Nishi Waseda, Shinjuku-ku, Tokyo 169, Japan	(03) 3203-4311	(03) 3205-1353
System Science Institute	3-4-1 Okuba, Shinjuku-ku, Tokyo 169, Japan	(03) 3200-2436	(03) 3232-7075
Kanagawa University	3-27-1 Rokkakubashi, Kanagawa-ku, Yokohama, Kanagawa 221-8686, Japan	(045) 481-5661	
Research Institute for Engineering	3-27-1 Rokkakubashi, Kanagawa-ku, Yokohama, Kanagawa 221-8686, Japan	(045) 481-5661	(045) 413-3643
Kanto Gakuin University	4834 Mutsuura-cho, Kanazawa-ku, Yokohama 236-8501, Japan	(045) 781-2001	

* The addresses and phone numbers listed for the universities are those of the secretariats of the graduate schools.
† From outside Japan, dial the international access code ("011" in the U.S.) + country code (81) + area code (omit "0"), then rest of number.

Japanese Private Universities and Their Attached Institutes and Laboratories (Science and Technology Only)	Address*	Phone Number†	Fax Number†
Institute of Science Technology	4834 Mutsuura-cho, Kanazawa-ku, Yokohama 236-8501, Japan	(045) 781-2001	(045) 784-8153
Osawa Memorial Institute of Architectural-Environmental Engineering	4834 Mutsuura-cho, Kanazawa-ku, Yokohama 236-8501, Japan	(045) 781-2001	
The Sanno Institute of Management	1573 Kamikasuya, Isehara, Kanagawa 259-1197, Japan	(0463) 92-2211	
Institute of Information Science	1573 Kamikasuya, Isehara, Kanagawa 259-1197, Japan	(0463) 92-2211	(0463) 93-0554
Shonan Institute of Technology	1-1-25, Nishi-Kaigan, Tsujido, Fujisawa, Kanagawa 251-8511, Japan	(0466) 34-4111	
Kanazawa Medical University	1-1 Daigaku, Uchinada-cho, Kawakita-gun, Ishikawa 920-0293, Japan	(076) 286-2211	
Institute of Human Genetics	1-1 Daigaku, Uchinada-cho, Kawakita-gun, Ishikawa 920-0293, Japan	(076) 286-2464	(0762) 86-2312
Institute for Tropical Medicine	1-1 Daigaku, Uchinada-cho, Kawakita-gun, Ishikawa 920-0293, Japan	(076) 286-2211	(0762) 86-0224
Kanazawa Institute of Technology	7-1 Ogigaoka, Nonoichi-machi, Ishikawa-gun, Ishikawa 921-8501, Japan	(076) 248-1100	
Fukui University of Technology	3-6-1 Gakuen, Fukui, Fukui 910-0028, Japan	(0776) 22-8111	
Institute of Industrial Research	3-6-1 Gakuen, Fukui, Fukui 910-0028, Japan	(0776) 22-8111	(0776) 22-8117
Aichi Medical University	21 Aza Karimata, Oaza Iwasaku, Nagakute-cho, Aichi-gun, Aichi 480-1195, Japan	(0561) 62-3311	
Radioisotope Research Center	21 Aza Karimata, Oaza Iwasaku, Nagakute-cho, Aichi-gun, Aichi 480-1195, Japan	(0561) 62-3311	
Physical Fitness Sports Medicine Rehabilitation Center	21 Aza Karimata, Oaza Iwasaku, Nagakute-cho, Aichi-gun, Aichi 480-1195, Japan	(0561) 62-3311	
Institute for Medical Science of Aging	21 Aza Karimata, Oaza Iwasaku, Nagakute-cho, Aichi-gun, Aichi 480-1195, Japan	(0561) 62-3311	(0561) 63-3531
Institute for Molecular Science of Medicine	21 Aza Karimata, Oaza Iwasaku, Nagakute-cho, Aichi-gun, Aichi 480-1195, Japan	(0561) 62-3311	(0561) 63-3532
Daido Institute of Technology	10-3 Takiharu-cho, Minami-ku, Nagoya, Aichi 457-8530, Japan	(052) 612-6111	
Materials Engineering Laboratory	2-21 Daido-cho, Minami-ku, Nagoya, Aichi 457, Japan	(052) 611-0513	(052) 612-5653

The Advanced Research Institute for Science and Engineering

Institution	Address	Phone	
Chubu University	1200 Matsumoto-cho, Kasugai, Aichi 487-8501, Japan	(0568) 51-1111	
Research Institute for Science and Technology	1200 Matsumoto-cho, Kasugai, Aichi 487-8501, Japan	(0568) 51-1111	(0568) 51-1141
Kyoto Sangyo University	Motoyama, Kamigamo, Kita-ku, Kyoto, Kyoto 603-8555, Japan	(075) 701-2151	
Institute of Computer Sciences	Motoyama, Kamigamo, Kita-ku, Kyoto, Kyoto 603-8555, Japan	(075) 701-2151	
Doshisha University	Karasuma Imadegawa Higshiiru, Kamigyo-ku, Kyoto, Kyoto 602-8580, Japan	(075) 251-3110	
Science and Engineering Research Institute	Karasuma Imadegawa Higshiiru, Kamigyo-ku, Kyoto, Kyoto 602-8580, Japan	(075) 251-3965	
Ritsumeikan University	56-1, Tojiin Kitamachi, Kita-ku, Kyoto, Kyoto 603-8577, Japan	(075) 465-1111	
Research Institute of Science and Engineering	56-1, Tojiin Kitamachi, Kita-ku, Kyoto, Kyoto 603-8577, Japan	(075) 465-1111	(075) 463-7475
Ryukoku University	67 Fukakusa-Tsukamotocho, Fushimi-ku, Kyoto, Kyoto 612-8577, Japan	(075) 642-1111	
Osaka Sangyo University	3-1-1 Nakagakiuchi, Daito, Osaka 574-8530, Japan	(0720) 75-3001	
The Institute for Industrial Research	3-1-1 Nakagakiuchi, Daito, Osaka 574-8530, Japan	(0720) 75-3001	
Osaka Dental University	8-1 Hanazono-cho, Kuzuha, Hirakata, Osaka 573-1121, Japan	(0720) 64-3111	
Research Institute for Clinical Dentistry	OMM Bldg, 1-7-31 Otemae, Chuo-ku, Osaka, Osaka 540, Japan	(06) 943-2234	
Kansai University	3-3-35 Yamate-cho, Suita, Osaka 564-8680, Japan	(06) 388-1121	
Research Institute of Industrial Technology	3-3-35 Yamate-cho, Suita, Osaka 564-8680, Japan	(06) 388-1121	
Kinki University	3-4-1 Kowakae, Higashi Osaka, Osaka 577-8502, Japan	(06) 721-2332	
Fisheries Laboratory	3153 Kogaura, Shirahama-cho, Nishimusai-gun, Wakayama 649-22, Japan	(0739) 42-2625	(0739) 42-2634
Atomic Energy Research Institute	3-4-1 Kowakae, Higashi Osaka, Osaka 577-8502, Japan	(06) 721-2332, ext. 4415	(06) 721-2353
Institute of Food Science	3-4-1 Kowakae, Higashi Osaka, Osaka 577-8502, Japan	(06) 721-2332	
Environmental Science Institute	3-4-1 Kowakae, Higashi Osaka, Osaka 577-8502, Japan	(06) 721-2332	
Research Institute for Science and Technology	3-4-1 Kowakae, Higashi Osaka, Osaka 577-8502, Japan	(06) 721-2332	
Life Science Research Institute	377-2 Ohno Higashi, Osaka Sayama-shi, Osaka 589, Japan	(0723) 66-0221	

* The addresses and phone numbers listed for the universities are those of the secretariats of the graduate schools.
† From outside Japan, dial the international access code ("011" in the U.S.) + country code (81) + area code (omit "0"), then rest of number.

Japanese Private Universities and Their Attached Institutes and Laboratories (Science and Technology Only)	Address*	Phone Number†	Fax Number†
Institute of Oriental Medicine	377-2 Ohno Higashi, Osaka Sayama-shi, Osaka 589, Japan	(0723) 66-0221	
Research Institute of Hypertension	377-2 Ohno Higashi, Osaka Sayama-shi, Osaka 589, Japan	(0723) 66-0221	(0723) 66-0206
Kwansei Gakuin University	1-155 Kamigahara Ichiban-cho, Nishinomiya, Hyogo 662-8501, Japan	(0798) 54-6100	
Information Processing Research Center	1-155 Kamigahara Ichiban-cho, Nishinomiya, Hyogo 662-8501, Japan	(0798) 53-6111, ext. 3200	(0798) 51-0913
Okayama University of Science	1-1 Ridai-cho, Okayama, Okayama 700-0005, Japan	(086) 252-3161	
Hiruzen Research Institute	Fukuda, Kawakami-mura, Maniwa-gun, Okayama 717-06, Japan	(0867) 66-3642	
Hiroshima University of Economics	37-1, 5-chome, Gion, Asaminami-ku, Hiroshima, Hiroshima 731-0192, Japan	(082) 871-1000	
Information Processing Center	37-1, 5-chome, Gion, Asaminami-ku, Hiroshima, Hiroshima 731-0192, Japan	(082) 878-3041	(082) 874-5635
Fukuyama University	985-1 Aza Sanzo, Higashimura-machi, Fukuyama, Hiroshima 729-0292, Japan	(0849) 36-2111	
Foundation for Industrial Science Research	985-1 Aza Sanzo, Higashimura-machi, Fukuyama, Hiroshima 729-0292, Japan	(0849) 36-2111	(0849) 36-2213
Kurume University	67 Asahi-cho, Kurume, Fukuoka 830-0011, Japan	(0942) 35-3311	
Institute of Cardiovascular Disease	67 Asahi-cho, Kurume, Fukuoka 830-0011, Japan	(0942) 35-3311	
Institute of Brain Diseases	67 Asahi-cho, Kurume, Fukuoka 830, Japan	(0942) 35-3311, ext. 505	(0942) 32-6278
University of Occupational and Environmental Health	1-1 Iogaoka, Yawata Nishi-ku, Kitakyushu, Fukuoka 807-8555, Japan	(093) 603-1611	
Institute of Industrial Ecological Science	1-1 Iogaoka, Yawata Nishi-ku, Kitakyushu, Fukuoka 807-8555, Japan	(093) 603-1611	(093) 692-1838
Fukuoka University	8-19-1 Nanakuma, Jonan-ku, Fukuoka, Fukuoka 814-0180, Japan	(092) 871-6631	
Central Research Institute	8-19-1 Nanakuma, Jonan-ku, Fukuoka, Fukuoka 814-0180, Japan	(092) 871-6631	(092) 862-4431
Nagasaki Institute of Applied Science	536 Amiba-machi, Nagasaki, Nagasaki 851-0193, Japan	(095) 839-3111	
The Institute of Regional Science	536 Amiba-machi, Nagasaki, Nagasaki 851-0193, Japan	(095) 839-3111	(095) 839-0584
Technical Research Laboratory	536 Amiba-machi, Nagasaki, Nagasaki 851-0193, Japan	(095) 839-3111	(095) 839-0584
Computer Science Center	536 Amiba-machi, Nagasaki, Nagasaki 851-0193, Japan	(095) 839-3111	(095) 839-0584

Appendix 4
Concise History of Japan

Year	Period or Era	Event
~10,000 B.C.	Paleolithic Stone Age	Aborigines resided in Japan.
		People migrated from Mongolia, China, Korea, and Southeast Asia, among other places.
660 BC	The first emperor, Jinmu*	
~2nd Century B.C.	Jomon Period	Hunting, fishing, and gathering were sources of livelihood.
~3rd Century A.D.	Yayoi Period	Rice cultivation was initiated and metal instruments were used.
100 A.D.		More than 100 small, independent countries existed throughout the Japanese archipelago.
4th Century A.D.	Asuka Period	Japan began to be consolidated by successive emperors.
538 (552†)		Buddhism was introduced into Japan from Kudara (Korea).
592~710		Empress Suiko acceded to the throne.
604		The first national constitution with 17 articles was enacted.
607		Horyuji was built (burned down in 670; rebuilt in the early 8th century).
630		The first mission was sent to T'ang (China) for a few years' study.
708		Wadokaiho (the cast copper coin) was put into use.
710~784	Nara Period	The capital (kyo) was transferred to Heijo-kyo (Nara).
752		Todaiji Temple and its statue of Buddha were constructed in Nara.

* According to the Kojiki and the Nihonshoki, written in the 8th century.
† According to another version.

Year	Period or Era	Event
784		The capital was transferred to Nagaoka-kyo near Kyoto.
794~1189	Heian Period	The capital was transferred to Heian-kyo (Kyoto).
1017	Era of noble Fujiwara (~1156)	Michinaga Fujiwara was appointed Minister of State by Emperor.
1167	Era of Taira family (~1185)	Kiyomori Taira was appointed Minister of State by Emperor.
1190~1297	Kamakura Period	Yoritomo Minamoto established the feudal government in Kamakura in 1183.
1192	(Beginning of Warriors Age)	Yoritomo Minamoto was appointed General (Shogun) by Emperor.
1274 and 1281	Mongolian attacks	Tokimune Hojo (successor of the government) defended the nation.
1333~1392	Nanbokucho Period	Emperor Godaigo overthrew the Kamakura government and restored it to Kyoto.
	1336 (Japan split between South and North Emperors	Takauji Ashikaga established the feudal government in Kyoto.
1398~1493	Muromachi Period	The emperorship was united in 1392 and Lord Ashikaga took power.
1467–1477		Rebellion Ohnin no Ran took place in Kyoto
1493		Masamoto Hosokawa expelled Yoshitane Ashikaga from Kyoto.
1493~1565	Sengoku Period	Warriors fought for the rule of the nation.
1543		The Portuguese brought guns to Japan in 1543
1549		Christianity was brought to Japan in 1549.
1568 (~1602)	Azuchi Momoyama Period	Nobunaga Oda governed Kyoto, but was killed in Honnoji rebellion in 1582.
1584		Hideyoshi Toyotomi entered Osaka Castle and was later appointed Minister of State.
1600		Tokugawa beat Toyotomi at Battle of Sekigahara.
1603~1867	Edo Period	Ieyasu Tokugawa set up the feudal government in Edo (Tokyo).
1633	(Final period of Warriors Age)	The National Isolation Order was enforced by Iemitsu Tokugawa.
1853	Resumption of external trade	Arrival of warships of Commodore Perry of the U.S. in Japan

Year	Period or Era	Event
1867~1911	Meiji Era (Emperor Mutsuhito)	Power was returned to Emperor from the Tokugawa Government in 1867.
1885		The cabinet system of Japan was inaugurated.
1889		The Imperial Constitution of Japan was promulgated.
1894 (~1895)		Japan won the Sino-Japanese War.
1904 (~1905)		Japan won the Russo-Japanese War.
1912~1926	Taisho Era	
1914 (~1918)	(Emperor Yoshihito)	Japan participated in World War I and contributed to the victory.
1926~1989	Showa Era	
Dec 1941~Aug 1945	(Emperor Hirohito)	Japan initiated the Pacific War (part of World War II) and lost to the allies.
1951		The U.S.–Japan Peace and Security Treaties were signed.
October 1964		The Olympic Games were held in Tokyo.
1970		International Exposition (Expo '70) was held in Osaka.
1973		The first oil crisis occurred. (The second occurred in 1979.)
1979		The Japan-China Peace and Comity Treaty was signed.
1989~	Heisei Era (Emperor Naruhito)	
1991		Bubble economy ended and economic slump started in Japan.
March 1999		Japan's Industry Competitiveness Council was established.

* According to the Kojiki and the Nihonshoki, written in the 8th century.
† According to another version.

Appendix 5
Japanese National Holidays

Day	Japanese National Holidays	
	In English	In Japanese (Roman Denotation)
Jan 1	New Year's Day	元日 (Ganjitsu)
Jan 10	Coming-of-Age Day	成人の日 (Seijin-no-hi)
Feb 11	National Foundation Day	建国記念日 (Kenkoku-kinen-bi)
(March)	Vernal Equinox Day	春分の日 (Shunbun-no-hi)
Apr 29	Greenery Day	緑の日 (Midori-no-hi)
	(formerly, Emperor Showa's Birthday)	(旧昭和天皇誕生日)
May 3	Constitution Memorial Day	憲法記念日 (Kenpou-kinen-bi)
May 4	National Holiday	国民の休日 (Kokumin-no-kyujitsu)
May 5	Children's Day	子供の日 (Kodomo-no-hi)
Jul 20	Marine Day	海の日 (Umi-no-hi)
Sep 15	Respect-for-the Aged Day	敬老の日 (Keirou-no-hi)
(September)	Autumnal Equinox Day	秋分の日 (Shubun-no-hi)
Oct 10	Health-Sports Day	体育の日 (Taiiku-no-hi)
	(In commemoration of the opening day of the Tokyo Olympic Games in 1964)	
Nov 3	Culture Day	文化の日 (Bunka-no-hi)
Nov 23	Labor Thanksgiving Day	勤労感謝の日 (Kinrou-kansha-no-hi)
Dec 23	Emperor's Birthday	天皇誕生日 (Tennou-tanjou-bi)

Appendix 6

Rental Fee of Apartments and Offices in Urban Areas in Kanto and Osaka Regions, Japan

Apartment

Area or Prefecture	Location: Ward (ku) or City	Average Monthly Rental Fee (yen per m²)		
		Jan 99	Jul 99	Feb 00
Tokyo Metropolitan Area	Chiyoda, Chuo, Minato	4,064	4,191	4,262
	Shinjuku, Suginami, Nakano	3,097	3,145	3,100
	Shinagawa, Ota, Meguro / Shibuya, Setagaya	3,114	3,137	3,146
	Toshima, Itabashi, Nerima / Kita, Bunkyo, Taito	2,700	2,683	2,677
	Arakawa, Adachi, Katsushika / Sumida, Koto, Edogawa	2,308	2,248	2,336
	Other than the above	2,425	2,326	2,312
Kanagawa Prefecture	Kawasaki, Northern Yokohama	2,235	2,219	2,236
	Central Yokohama	2,554	2,559	2,567
	Western Yokohama	2,068	2,068	2,090
	Southern Yokohama	2,020	1,906	1,997
	Fujisawa, Kamakura, Zushi, Yokosuka, Miura	1,865	1,835	1,855
	Sagamihara, Zama, Ebina, Atsugi, Isehara	1,730	1,651	1,651
Chiba Prefecture	Matsudo, Kashiwa, Abiko / Nagareyama, Noda, Kamagaya	1,591	1,499	1,583
	Ichikawa, Urayasu, Funabashi / Narashino, Yachiyo	1,886	1,774	1,814
	Chiba, Yotsukaido, Sakura / Ichihara, Sodegaura	1,548	1,594	1,567
Saitama Prefecture	Tokorozawa, Sayama, Iruma	1,690	1,564	1,619
	Omiya, Yono, Urawa, Iwatsuki	1,774	1,735	1,785
	Kawaguchi, Hatogaya, Warabi, Toda / Wako, Asaka, Niiza	1,920	1,852	1,830

Office

Location: Ward or City	Town	Average Monthly Fee	
		Aug 99	Feb 2000
Chiyoda-ku Tokyo	Kanda, Ochanomizu	6,634	6,099
	Akihabara, Iwamoto-cho	5,256	4,928
	Suidobashi, Iidabashi, Kudan	5,735	5,434
	Kojimachi, Bancho	6,742	6,563
Chuo-ku Tokyo	Yaesu, Kyobashi, Nihonbashi	7,975	7,736
	Kotenma-cho, Horidome, Ningyo-cho	4,832	4,735
	Ginza	7,809	7,322
	Hacchobori, Kayaba-cho, Sinkawa	5,976	5,278
	Shinbashi, Toranomon	6,990	6,407
Minato-ku Tokyo	Akasaka, Aoyama	6,808	6,292
	Hamamatsu-cho, Tamachi, Shinagawa	5,640	5,993
	Kaigan, Shibaura, Kounan	4,944	5,292
Shinjuku-ku Tokyo	Nishi-Shinjuku	7,417	6,633
	Shinjuku	5,244	5,278
Shibuya-ku Tokyo	Shibuya	6,421	6,243
	Ebisu, Hiroo	6,500	5,683
Yokohama	Kannai	3,821	3,766
	In the Vicinity of Yokohama Station	3,730	4,708
	Shin-Yokohama	4,830	3,719
Osaka	Umeda	5,410	5,345
	Minamimori-machi	4,160	4,059
	Yodoyabashi, Honmachi	4,414	4,369
	Senba	4,189	4,095
	Shinsaibashi, Nanba	5,197	5,150
	Shin-Osaka	4,170	4,066

Source: At Home, Ltd., Ota-ku, Tokyo, Japan for Apartment, Miki-Shoji, Ltd., Chuo-ku, Tokyo for Office

Appendix 7
JR Maps of Japan and Greater Tokyo

The following are non-JR lines (and sections), indicated as ✳1- ✳7 in the map, that may be traveled on (though the JAPAN RAIL PASS is not valid for them) by JR-line exp limited express trains: ✳1 Hokuetsu Kyuko (Muikamachi-Saigata); ✳2 Izu Kyuko (Ito-Izukyu Shimoda); ✳3 Noto Railway (Wakura Onsen-Vajima/Takojima); ✳4 Ise (Kawarada-Tsu); ✳5 Kitakinki Tango Railway (Nishimaizuru-Toyooka; Fukyuchiyama-Miyazu); ✳6 Chizu Kyuko (Kamigori-Chizu); ✳7 Tosa Kurishio Railway (Kubokawa-

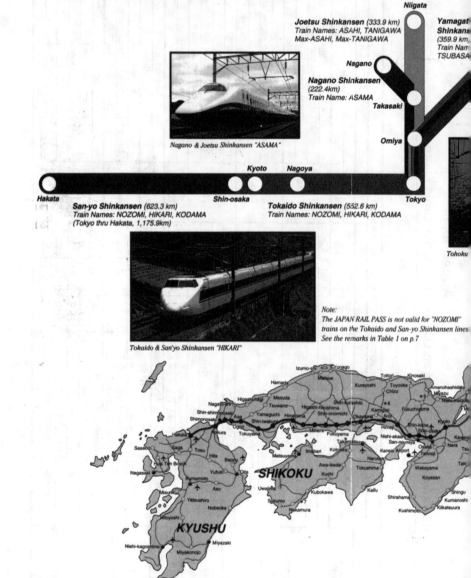

Shinkansen Super Express

Joetsu Shinkansen (333.9 km)
Train Names: ASAHI, TANIGAWA
Max-ASAHI, Max-TANIGAWA

**Yamagat
Shinkans**
(359.9 km,
Train Nam
TSUBASA

Niigata

Nagano

Nagano Shinkansen
(222.4km)
Train Name: ASAMA

Takasaki

Omiya

Nagano & Joetsu Shinkansen "ASAMA"

Kyoto Nagoya

Hakata

Shin-osaka

Tokyo

San-yo Shinkansen (623.3 km)
Train Names: NOZOMI, HIKARI, KODAMA
(Tokyo thru Hakata, 1,175.9km)

Tokaido Shinkansen (552.6 km)
Train Names: NOZOMI, HIKARI, KODAMA

Tohoku

Note:
The JAPAN RAIL PASS is not valid for "NOZOMI"
trains on the Tokaido and San-yo Shinkansen lines
See the remarks in Table 1 on p.7

Tokaido & San'yo Shinkansen "HIKARI"

SHIKOKU

KYUSHU

Akita Shinkansen "KOMACHI"

HOKKAIDO

HONSHU

(535.3 km)
KO, NASUNO
NASUNO

Morioka

● **Stations in Tokyo area**
that have JAPAN RAIL PASS exchange offices

Ikebukuro

Ueno

Shinjuku

Tokyo

Shibuya

Airport
Terminal 2

Shinagawa

Narita Airport
(Terminal 1)

Yokohama

Narita Airport-Tokyo
53 min. (shortest) by JR's limited express "Narita Express"
About 90 min. by JR's rapid train "Airport Narita"
About 80 min. by limousine bus
About 60 min. by Keisei Railways' "Skyliner"
(Narita Airport - Keisei Ueno Station)

● **Stations in Kansai area**
that have JAPAN RAIL PASS exchange offices

Shin-osaka

Kyoto

Osaka

JR Nanba

Kyobashi

Kansai Airport

Tennoji

Hineno

Kansai Airport - Shin'osaka/Osaka/Kyoto
48 min. (shortest) to Shin'osaka, 73 min. (shortest) to
Kyoto by JR's limited express "Kansai Airport Express
Haruka"
About 55 min. to Osaka by Kansai Airport Special Rapid
Service

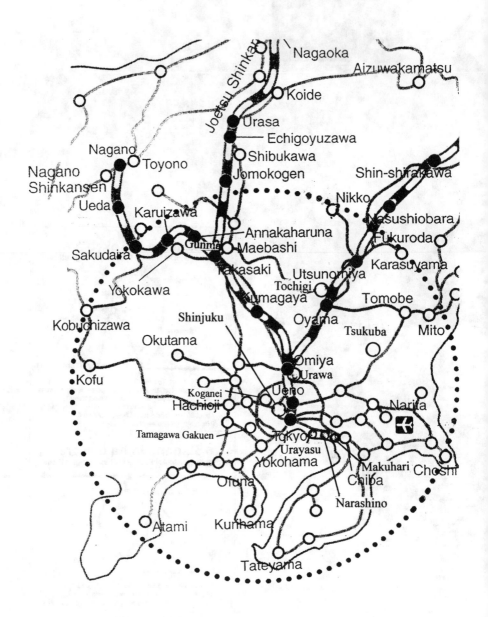

Appendix 8
Japan and Surrounding Pacific Rim Economies

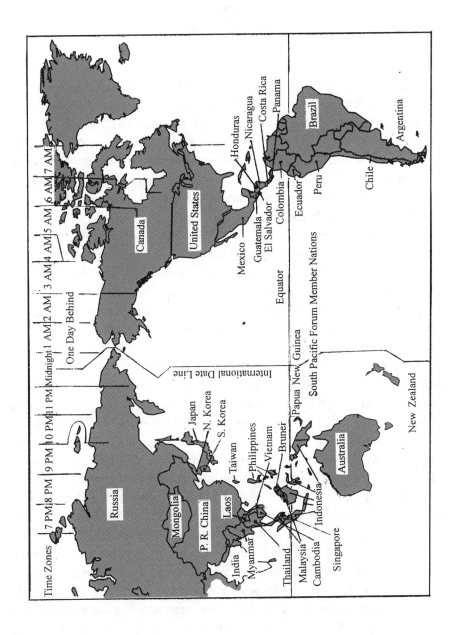

Activities of the ASME Japan Section

Junjiro Iwamoto
Chair (1998–2000)

ASME Japan was established in 1986 by engineers in Japan who were earnestly involved in the activities of ASME. The founding chairman was the late Professor Ichiro Watanabe. In 1990, ASME Japan was officially recognized by ASME as its Japan Chapter. Regular publication of the Newsletter and frequent meetings and other events began to occur around this time. But even before the formal approval of the International chapter, the Newsletter was being published, and some other activities took place as well.

On March 30, 1990, the annual meeting of ASME Japan was held, and the first Chair, Professor Wataru Nakayama, and Board members of the newly approved Japan Chapter were elected. Then-President of ASME Professor Arthur Bergles was invited to the meeting and gave a talk entitled "ASME in the 1990s," in which he explained the organization and activities of ASME. After the talk, attendees took a tour of Sony's media world, and then a dinner reception was given. In the following two years, seminars, short courses, and lecture meetings were planned and held under the leadership of chair and board members.

Since then, chair and board members have been elected every two years. At each annual meeting and at other meetings, prominent guest speakers have been invited and have given seminars-among them Professor Akira Sakurai, Melville Medal recipient; Professor Shoichi Furuhama, President of the Musashi Institute of Technology; Professor M. Ragsdell, University of Missouri; and Dr. Edward M. Malloy, Counselor for Scientific & Technological affairs, U.S. Embassy. The content of the talks ranged from general interests to technological problems. In 1996, the Japan Chapter became the Japan Section in Region XIII of ASME International. In the first half of 1999 alone, we had two general meetings. At one of the meetings, held on March 26, 1999, Dr. W. Nakayama, President of ThermTec International and a former chair of ASME Japan, gave a talk entitled "Japan in Transition: Impacts of Industrial Restructuring on the Engineering Profession." He talked about the current economic situation and related problems facing the engineering profession in Japan. And he spoke of ways ASME Japan can contribute to improving the present trying situation of the engineering profession in Japan.

At the annual meeting held on June 25, 1999, Messrs. Kazuhiko Watanabe, General Manager, and Yoshinori Kajimura, Manager, of the High Pressure Gas Safety Institute of Japan, gave a very interesting talk, "The New Era of Mega-Competition of Codes and Standards for Boiler and Pressure Vessels." They talked about the emergence of a new era of mega-competition for codes and standards, and the current situation. On December 8, 1999, we hosted a meeting for a lecture entitled "Why and How to Deal with Uncertainties," by Dr. Isaac Elishakoff, Professor at Florida Atlantic University, and held a dinner reception in his honor. The meetings and talks were well attended and received.

For seven years the Newsletter of ASME Japan had not been published, but in June 1999 publication resumed, and from now on publication twice or thrice a year is planned.

In 1987, in the early days of ASME Japan, membership stood at 285, but now it has grown to more than 700 and is still growing. Now we have a larger membership than any other section outside North America. ASME Japan will continue to be more active for the benefit of its members.

About the Contributors

Kathryn B. Aberle is the Associate Executive Director of the Accreditation Board for Engineering and Technology (ABET), responsible for the management and supervision of ABET headquarters operations. She is the staff liaison to ABET's Educational Policy and Finance Committees, and previously served as a liaison to the International Activities Committees, and an ex officio member of ABET's Strategic Planning Committee. She also serves as ABET's primary liaison to the U.S. Department of Education, the Council on Higher Education Accreditation (CHIEA), and other governmental and nongovernmental organizations involved in accreditation. From 1989 to 1995, Ms. Aberle was at the Society of Petroleum Engineers (SPE), an international organization. Ms. Aberle is a Certified Association Executive (CAE). She earned her B.A. from the University of California at Riverside and her MBA from the University of Texas at Austin.

Michael W. Barnett is an American who is an assistant professor in the Department of Civil Engineering and Engineering Mechanics at McMaster University, Hamilton, Ontario. Prior to holding this position, he was a lecturer and a postdoctoral researcher in the Department of Environmental Engineering, Kyoto University. He originally came to Japan on a Monbusho scholarship and subsequently was hired as the first foreign faculty member in his department at Kyoto University. He holds a B.A. in psychology and an M.S. in environmental engineering, both from the University of Cincinnati, and a Ph.D. in environmental engineering from Rice University, Houston.

Daniel K. Day is an English teacher and a freelance translator (Japanese to English). His work includes weekly columns for the *Asahi Evening News*. He holds a B.S. in mechanical engineering from the University of Colorado (Boulder). After working for Boeing Commercial Aircraft Corporation for two and a half years, he moved to Sapporo, Japan, in 1981 and stayed there for 15 years.

Robert M. Deiters, an American who has lived in Japan since 1952, is a professor in the Electrical-Electronics Engineering Department of Sophia University in central Tokyo. He is a member of the Society of Jesus and an ordained Catholic priest. He holds degrees in theology from Sophia University and a doctorate in engineering (kogaku hakushi) from the University of Tokyo. At

Sophia University he has been director of the Computer Center and dean of the Faculty of Science and Technology. At present he teaches and does research in computer networks and computer applications in engineering.

Shuichi Fukuda is a professor of systems engineering in the Management Engineering Department and advisor to the President of the Tokyo Metropolitan Institute of Technology. He served as associate professor at the Welding Research Institute of Osaka University and taught in the department of Precision Machinery Engineering, University of Tokyo. He holds bachelor's, master's, and doctor's degrees (kogaku hakushi) in mechanical engineering from the University of Tokyo.

Stephen A. Hann is an American who moved to Japan in 1989. He was raised in Virginia Beach, Virginia, and holds degrees in applied mechanics from Old Dominion University and in mechanical engineering from the University of Michigan. He and his wife, the late **Deborah A. Coleman Hann**, who was raised in Toledo, Ohio, established the Mechanical Simulation Corporation in Ann Arbor, Michigan, and moved to Japan in March 1989 to establish an office in Tokyo and stayed there until 1994.

Charles R. Heidengren is an American consulting engineer. He holds a degree in civil engineering from the Cooper Union School of Engineering, with graduate studies in solid mechanics at Columbia University. He was manager for civil engineering in the engineering and construction division of Kawasaki Steel Corporation. Prior to this, he spent five years as senior technical adviser and project manager for Pacific Consultants International. He has more than 35 years' experience in geotechnical and foundation engineering, with additional experience in project management and business development since 1979. He is a licensed professional engineer and was founder and past president of the Japan Section of the American Society of Civil Engineers.

Hiroshi Honda, born in Kanazawa City, earned his M.S. in engineering mechanics from the Pennsylvania State University in 1976 and his bachelor's and doctor's degrees in mechanical engineering (with theses in the areas of strength design of gears and offshore structure, fracture mechanics and fatigue) from Kyoto University. He also holds P.E. licenses from the states of Minnesota and Texas. He has served in private industry in Tokyo in the roles of member of the engineering staff at machinery headquarters; chief research engineer; associate manager for corporate planning, business planning, and marketing departments; manager for an environment and energy department; and general manager for an engineering center. He has also worked for the Institute of Energy Economics, Japan; the Japanese Committee for Pacific Coal Flow (JAPAC); the Japan Microgravity Center; and the Japan Space Utilization Promotion Center (JSUP). Dr. Honda served as chairman and secretary for various committees of ASME Japan, JSME, JSUP, JAPAC, and others—such as editorial, research planning, research on industry frontiers, investigation on space utilization, energy master plan and coal flow committees—and authored and coauthored more than 110 papers, articles, reports, and books in both English and Japanese. He received the distinguished alumnus award from Penn State in 1998, and established Honda International, Ltd. (7-2-710, Yatsu 6-Chome, Narashino City, Chiba 275-0026, Japan; telephone and fax numbers: 81 + (0)47 477-8571; e-mail address: Hondah9876@aol.com), in April 1999. He has a continuing interest in the activities of international organizations such as APEC, OECD, and the U.N., with which he was actively involved, serving as a lecturer, chair, reporter, and organizer, in his previous positions in Japan. He served as president of Japan Chapter of Penn State Alumni Association and is a member of Sokyu Club (Old Boy Club of Kyoto University Soft Tennis Team), Narashino International Association (NIA), ASME and JSME.

Junjiro Iwamoto is a professor of mechanical engineering at Tokyo Denki University. He received bachelor's, master's, and doctor's degrees (kogaku hakushi) from Keio University. He serves as chairman of the ASME Japan Section for the 1998–2000 term.

Robert Latorre is a professor of naval architecture and marine engineering at the University of New Orleans. He completed his B.S.E. at the University of Michigan. He studied Japanese at Middlebury College in Vermont and then completed an M.S.E. and Dr. Eng (Ph.D.) at the University of Tokyo. In 1986–87 he taught fluid mechanics in the Mechanical Engineering Department of the University of Tokyo.

Peter E. D. Morgan is an American (formerly British) living in Thousand Oaks, California, where he works at Rockwell International Science Center. He has a Ph.D. in inorganic chemistry from Imperial College, London University, and is a fellow of the American Ceramic Society. He has visited Japan numerous times, including several stays of a few months at the Superconductivity Center in Hitachi, at Osaka University, and at the National Industrial Research Institute in Nagoya (NIRIN).

Giuseppe Pezzotti is associate professor, Department of Materials, at Kyoto Institute of Technology (KIT). He was born in Rome and completed his diploma di laurea in mechanical engineering at Rome University and his doctor of engineering (Ph.D.) in materials science and engineering at Osaka University. He was with Osaka University, Tohoku University, and Toyohashi University of Technology prior to coming to KIT in 1996.

George D. Peterson is the Executive Director of the Accreditation Board for Engineering and Technology (ABET). He is the former Section Head of Faculty and Teacher Development, Division of Undergraduate Education, at the National Science Foundation (NSF) in Washington, D.C., and from 1988 until 1989 he served at the NSF as Program Director in the Undergraduate Science, Engineering and Mathematics Education Division. Dr. Peterson was Chairman of the Department of Electrical Engineering at the U.S. Naval Academy, Annapolis, Maryland, from 1983 to 1988. He was Assistant Vice President for Academic Affairs and professor of Electrical Engineering at Morgan State University in Baltimore, Maryland, from 1988 to 1993. Dr. Peterson is a Fellow of ABET and a Fellow of the Institution of Engineers of Ireland and is a licensed Professional Engineer in the states of Colorado and Maryland.

Winfred M. Phillips is Vice President for Research and Dean of the Graduate School at the University of Florida. He earned his B.S.ME degree from Virginia Polytechnic Institute in Mechanical Engineering and his M.A.E. and D.Sc. degrees in Aerospace Engineering from the University of Virginia. Dr. Phillips formerly served as Dean of Engineering and Associate Vice President at the University of Florida, and as professor of mechanical engineering and head of the School of Mechanical Engineering at Purdue University. He also served as assistant, associate, and full professor of aerospace engineering, as Associate Dean for Research at the College of Engineering, and as Acting Chairman for the Intercollegiate Bioengineering Program at the Pennsylvania State University; and as Visiting Professor at the University of Paris (hemodynamics research). He is a Fellow of the American Society of Mechanical Engineers, and has served as President of ASME, President of the Accreditation Board for Engineering and Technology (ABET), and President of the American Society for Engineering Education. Author and coauthor of over 150 publications, Dr. Phillips has presented over 280 technical presentations, lectures by invitation, and seminars. His personal research and teaching interests include mechanical engineering, fluid mechanics, and biomedical engineering.

Kazuo Takaiwa (now deceased) was a Japanese, and a representative of ITK, Inc. (Internationale Technik und Konsulent Inc.). He was a consultant for project engineering, construction, maintenance, training of company staff, and cross-cultural issues. He held a bachelor's degree from Kyushu University and a doctor's degree (kogaku hakushi) in plant construction control systems from the University of Tokyo. He was a director, project manager, and construction manager for overseas projects and a pressure vessel factory manager at Chiyoda Corporation.

Craig Van Degrift is an American physicist in the Electricity Division of the National Institute of Standards and Technology carrying out research on the quantum Hall effect in support of the U.S. standard of electrical resistance. A native of Los Angeles, he received his B.S. in physics from Harvey Mudd College in 1966, an M.A. from the University of California at Irvine in 1967, and a Ph.D. from Irvine in 1974. He was able to combine his professional physics activity with a long-term side interest in the Japanese language by visiting the Japanese Electrotechnical Laboratory for a year to participate in the Japanese quantum Hall research effort.

Raymond C. Vonderau is an American who was the regional vice president of the Tokyo office of Beloit Corporation (Beloit Nippon, Ltd.) He received a B.S.ME degree from Purdue University in 1949 and joined Beloit in 1959. He has held various executive engineering positions for Beloit and its affiliate companies and has contributed to their success in the pulp and paper industry in North America. He is a member of the Technical Association of the Pulp and Paper Industry (TAPPI) and of ASME, and is a registered professional engineer in the state of Wisconsin. In 1995, Mr. Vonderau retired from active work in the Pulp and Paper Industry and is now living in Palm Desert, California.

Dimitrios C. Xyloyiannis, a Greek, worked as a project manager and engineer for Sandoz Industrial Technology, Ltd., headquartered in Switzerland. He is involved in two projects in Japan. He is a professional marine engineer and holds a master's degree in mechanical engineering. He is a full member of the Greek Management Society and the Greek Operational Research Society (GORS). He was a lecturer in technical seminars offered by GORS until 1989, and is a member of ASME.

Joyce Yamamoto, a Japanese-American materials scientist, initially came to a Japanese national laboratory on a postdoctoral fellowship. At the end of the fellowship she worked for a private company as a visiting researcher. She received a B.S. in ceramic engineering from the University of Illinois, Urbana-Champaign, and an M.S. in ceramic science and a Ph.D. in materials science from the Pennsylvania State University. Returning to the United States after three years in Japan, she was a staff researcher at Cornell University. Since that time she has worked for Motorola in research and development.

Glossary of Japanese Words

ba	Place and frame for one's life and activity. (24, 85, 87)
bucho	Department general manager. (34, 35)
bujicho	Department deputy general manager. (34, 35)
chiho	Region or rural area. (114)
chodendo	Superconductivity. (170)
chuo kenkyusho	Central research laboratory. (169)
daigaku	University. (82)
dan-dan	Step by step. (172)
Edo	Former name of Tokyo used until 1867. (56, 59)
Edokko	Official designation for persons whose families have lived in Tokyo for three generations or longer. (60)
fukoku kyouhei	A slogan, "toward a wealthy and strongly armed nation," used by the Japanese government in the Meiji era. (90)
fusuma	Japanese paper door. (131)
gaijin	Foreigner. (161, 187, 193, 194)
gakubucho	Dean. (35, 36, 37)
gijutsushi	Registered consulting engineer. (162)
gonin-gumi	Five-person team through which families watched each other's behavior, under the social system that evolved during the Edo period. (57)
goui	Consensus. (177)
gyoza	Chinese fried meat and vegetables wrapped with a flour coating. (66)
hakushi	Ph.D. (162)
Handai	An abbreviation for Osaka (Han) University (Dai). (175)
haragei	The art (*gei*) of the belly (*hara*), where the "belly" signifies one's heart, what one is really thinking. The art is in transmitting one's true intention without putting it directly into words. (110)
harakiri (seppuku)	Committing suicide by cutting the belly with a sword (to restore samurai's honor). (57)

hara o yomu	To read another's intention or mind. (110)
hiragana	Kind of kana used for genuine Japanese words. (113, 114, 127, 128)
honne	One's true intention or mind. (110)
Honshu	Japan's main island. (60, 279)
Horyuji	The oldest existing wooden (Buddhist temple) building, originally established in 607, burned down in 670, and rebuilt in the early eighth century in Nara, Japan. (179)
jinmyaku	Human networking. (177)
jokyoju	Associate professor. (36, 37, 38)
Joochi Daigaku	Sophia University, Tokyo, Japan (82)
joshu	Research associate. (36, 37, 38)
JR	Japan Railway (a group of companies privatized from a national railway company). (114, 278, 280)
kabushiki gaisha	Joint-stock company. (213, 214)
kacho	Section manager. (34, 35)
Kagaku-Gijutsu Dantai Rengo	The United Science and Technology Organizations (Foundations) in Japan. (153)
kaisha	Company or corporation. (156)
kaizen	Improvement, implying many small, never ending steps. (172)
kajicho	Section deputy manager (34, 35)
kakaricho	Assistant manager. (34, 35)
kamikaze	"The wind of the gods," applied to a typhoon that stopped a Mongolian force attempting to invade Japan in the thirteenth century. (133)
kana	Japanese syllabic character; closest counterpart of alphabetic characters in English. (112, 113, 114)
kanji	Chinese character, used in the Japanese language to signify the meaning of a word, not its sound. (112, 113, 114, 127, 128, 129)
Kansai	Kyoto-Osaka-Koba-Nara area. (79)
Kansai-Chukyo	Kyoto-Osaka-Kobenara-Nagoya area. (79)
katakana	A kind of kana used for words of western origin. (114, 115, 127)
keiretsu	Conglomerate family or company group. (76, 77, 177)

ken	Prefecture. (167)
Kenkyusho	Research laboratory (169)
kikkake	Chance or opportunity or beginning or clue (77)
kocho	School principal (refers to high school principals for salary tables). (36, 37, 38)
kogaku hakushi	Doctor's degree in engineering, typically equivalent to German Dr.-Ing., conferred by Japanese universities, which certifies that a person has attained significant achievement in engineering research and has passed a qualifying examination. (Japanese universities adopted the Ph.D., or Hakushi, system in the 1990s.) (162, 283)
kogakushi	Bachelor's degree in engineering conferred by Japanese universities. (Gakushi means bachelor's degree.)
kogaku shushi	Master's degree in engineering conferred by Japanese universities. (Shushi means master's degree.)
kokoro	Soul or spirit. (174)
kokusai	International (169)
koohai	Younger person in "vertical" human relations. (85)
koushi	University lecturer. (36, 37, 38)
kyoju	Professor. (36, 37, 38)
kyotou	School vice principal (refers to high school vice principals for salary tables.) (36, 37, 38)
kyoushi	School teacher (refers to high school vice principals for salary tables). (36, 37, 38)
Marunouchi	Established business district in front of Tokyo station. (59)
Meiji-jidai	Meiji era, the reign of Emperor Mutsuhito (1867-1911). (58, 179)
Monbusho	Ministry of Education, Science, Sports and Culture. (202, 203, 208)
naniwabushi	Old Edo ballad that originated in Osaka. (77)
nanushi	Village headman in feudal periods in Japan. (57)
nemawashi	Literally, "root binding," or gaining a consensus of the necessary people. (66, 74, 111)
Nihon-go	The Japanese language. (113)
nominucation	A slang combination of the Japanese word *nomi*, "drinking," and the English word *communication*. (66)

romaji	Roman alphabet. (111)
sake	Rice wine. (173, 180)
sakura	Cherry blossom. (174)
samurai	Social rank in feudal periods in Japan equivalent to knight. (57, 179)
sempai	Older person in "vertical" human relations. (85)
Sendo Kenkyu	Leading research promoted by Japanese governmental organizations such as the Ministry of International Trade and Industry (MITI). (43)
senseisho	An oath form. (91)
shi	City. (167)
Shinto	The official religion of Japan incorporating the worship of ancestors and nature spirits (56)
shoji	Sliding paper door. (102)
shosha	Commodity trading company. (8, 186)
tabi	Traditional Japanese-style socks. (138)
tatami	Woven straw mats regularly used in Japanese houses. (102, 140)
tatemae	The way something is supposed to be, in contrast with *honne*. (110)
Tokubetsu Kenkyu Shoreihi	Monbusho's grant-in-aid for Japan Society for the Promotion of Science (JSPS) fellows. (417)
uchi-soto	Insider (uchi) and outsider (soto); us and them, or your group and competing group. (173, 174)
wa	Peace, harmony, and cooperation. (134)
yakuin	Executive, such as director, managing director, vice president, president, or chairman. (30)
yokozuna	The highest rank (grand champion) in sumo wrestling (68)
yuugen gaisha	Limited company. (213)

Index